T0335771

Springer Theses

Recognizing Outstanding Ph.D. Research

Aims and Scope

The series "Springer Theses" brings together a selection of the very best Ph.D. theses from around the world and across the physical sciences. Nominated and endorsed by two recognized specialists, each published volume has been selected for its scientific excellence and the high impact of its contents for the pertinent field of research. For greater accessibility to non-specialists, the published versions include an extended introduction, as well as a foreword by the student's supervisor explaining the special relevance of the work for the field. As a whole, the series will provide a valuable resource both for newcomers to the research fields described, and for other scientists seeking detailed background information on special questions. Finally, it provides an accredited documentation of the valuable contributions made by today's younger generation of scientists.

Theses are accepted into the series by invited nomination only and must fulfill all of the following criteria

- They must be written in good English.
- The topic should fall within the confines of Chemistry, Physics, Earth Sciences, Engineering and related interdisciplinary fields such as Materials, Nanoscience, Chemical Engineering, Complex Systems and Biophysics.
- The work reported in the thesis must represent a significant scientific advance.
- If the thesis includes previously published material, permission to reproduce this must be gained from the respective copyright holder.
- They must have been examined and passed during the 12 months prior to nomination.
- Each thesis should include a foreword by the supervisor outlining the significance of its content.
- The theses should have a clearly defined structure including an introduction accessible to scientists not expert in that particular field.

More information about this series at http://www.springer.com/series/8790

Huahua Xiao

Experimental and Numerical Study of Dynamics of Premixed Hydrogen-Air Flames Propagating in Ducts

Doctoral Thesis accepted by
University of Science and Technology of China,
Hefei, China

 Springer

Author
Dr. Huahua Xiao
State Key Laboratory of Fire Science
University of Science and Technology
   of China
Hefei
China

Supervisor
Prof. Jinhua Sun
University of Science and Technology
   of China
Hefei
China

ISSN 2190-5053   ISSN 2190-5061   (electronic)
Springer Theses
ISBN 978-3-662-48377-0   ISBN 978-3-662-48379-4   (eBook)
DOI 10.1007/978-3-662-48379-4

Library of Congress Control Number: 2015949445

Springer Heidelberg New York Dordrecht London
© Springer-Verlag Berlin Heidelberg 2016
This work is subject to copyright. All rights are reserved by the Publisher, whether the whole or part
of the material is concerned, specifically the rights of translation, reprinting, reuse of illustrations,
recitation, broadcasting, reproduction on microfilms or in any other physical way, and transmission
or information storage and retrieval, electronic adaptation, computer software, or by similar or
dissimilar methodology now known or hereafter developed.
The use of general descriptive names, registered names, trademarks, service marks, etc. in this
publication does not imply, even in the absence of a specific statement, that such names are exempt
from the relevant protective laws and regulations and therefore free for general use.
The publisher, the authors and the editors are safe to assume that the advice and information in this
book are believed to be true and accurate at the date of publication. Neither the publisher nor the
authors or the editors give a warranty, express or implied, with respect to the material contained
herein or for any errors or omissions that may have been made.

Printed on acid-free paper

Springer-Verlag GmbH Berlin Heidelberg is part of Springer Science+Business Media
(www.springer.com)

Parts of this thesis have been published in the following articles:

Xiao H, Sun J, Chen P (2014) Experimental and numerical study of premixed hydrogen/air flame propagating in a combustion chamber. J Hazard Mater 268:132–139 (Reproduced with Permission).

Xiao H, Wang Q, Shen X, An W, Duan Q, Sun J (2014) An experimental study of premixed hydrogen/air flame propagation in a partially open duct. Int J Hydrogen Energy 39:6233–6241 (Reproduced with Permission).

Xiao H, He X, Duan Q, Luo X, Sun J (2014) An investigation of premixed flame propagation in a closed combustion duct with a 90° bend. Appl Energy 134:248–256.

Xiao H, Wang Q, Shen X, Guo S, Sun J (2013) An experimental study of distorted tulip flame formation in a closed duct. Combust Flame 160:1725–1728 (Reproduced with Permission).

Xiao H, An W, Duan Q, Sun J (2013) Dynamics of premixed hydrogen/air flame in a closed combustion vessel. Int J Hydrogen Energy 38:12856–12864 (Reproduced with Permission).

Xiao H, He X, Wang Q, Sun Q (2013) Experimental and numerical study of premixed flame propagation in a closed duct with a 90° curved section. Int J Heat Mass Transfer 66:818–822.

Xiao H, Makarov D, Sun J, Molkov V (2012) Experimental and numerical investigation of premixed flame propagation with distorted tulip shape in a closed duct. Combust Flame 159:1523–1538 (Reproduced with Permission).

Xiao H, Shen X, Sun J (2012) Experimental study and three-dimensional simulation of premixed hydrogen/air flame propagation in a closed duct. Int J Hydrogen Energy 37:11466–11473 (Reproduced with Permission).

Xiao H, Wang Q, He X, Sun J (2011) Experimental study on the behaviors and shape changes of premixed hydrogen-air flames propagating in horizontal duct. Int J Hydrogen Energy 36:6325–6336 (Reproduced with Permission).

Xiao H, Wang Q, Sun J, He X, Yao L (2010) Experimental and numerical study on premixed hydrogen/air flame propagation in a horizontal rectangular closed duct. Int J Hydrogen Energy 35:1367–1376 (Reproduced with Permission).

Supervisor's Foreword

Understanding of the dynamics of premixed flames evolving in tubes is of great importance in a broad range of practical combustion phenomena of scientific and engineering interest, such as gas explosions in confined regions, and burning processes in internal combustion engines. A transient premixed flame in a tube is an extremely complex, dynamic process involving chemical kinetics, heat and mass transfer, and fluid dynamics. The complexity of the flame dynamics presents enormous challenges which combustion scientists and engineers have to face.

Great interest in new alternative fuels is generated associated with economic and environmental concerns in the use of fossil fuels, e.g., high fuel price, global warming and environmental pollution. Hydrogen as an energy carrier is one of the promising alternative fuels because of its potentially high efficiency and ultra-low harmful emissions. However, there are serious problems to overcome when using hydrogen as an energy carrier. The major issues in relation to combustion and safety are the unique characteristics of hydrogen due to its high diffusivity and reactivity, which can lead to leak, fire, and explosion hazards.

In the past eight years, Dr. Huahua Xiao has been devoted to premixed flame propagation in confined regions. In the present work, dynamics of premixed hydrogen–air flames propagating in ducts under a large variety of conditions are experimentally and numerically studied. A subsequent theoretical analysis is performed. Numerical methods and combustion models are suggested and validated against experimental and theoretical results. Important knowledge of premixed flame dynamics and combustion modeling is presented. Particularly, one outstanding finding proposed by Dr. Xiao is the "distorted tulip flame". Furthermore, the interactions between the flames and various physical phenomena, e.g., the combustion-generated flow, pressure wave and boundary layer, are examined to elucidate the underlying mechanisms that control the combustion processes.

Overall, this work provides us with important new insights and deeper understanding of premixed flame dynamics, and combustion of hydrogen–air premixtures in ducts.

Hefei   Prof. Jinhua Sun
July 2015

Abstract

Premixed combustion of chemically reactive gas mixture is a very fundamental subject for a broad range of issues of scientific and engineering interest, e.g., accidental explosions and propulsion applications. The fundamental understanding of premixed flame propagation phenomena is essential for the development of novel analytical and numerical combustion models. Premixed flame dynamics in confined vessels is of particular importance since it provides understanding of the burning processes taking place in internal combustion engines, and explains the mechanisms behind flame acceleration that can lead to transition from deflagration to detonation. In addition, hydrogen is a promising alternative energy carrier in the future, and it is of great importance to characterize the combustion behavior of its blends with air. Meantime, the development and the validation of contemporary combustion models of wide applicability are important for both hydrogen combustion applications and explosion safety.

This study aims to provide fundamental and in-depth investigation of premixed combustion and reliable prediction approaches for hydrogen combustion and explosion in air. Two primary aims are planned to be achieved in the present work. The first objective is to study the premixed combustion dynamics in tubes, i.e., flame propagation and pressure build-up, and explain the mechanisms underlying the dynamics of the premixed flame. Another important target of this study is to investigate hydrogen gas explosions in tubes, and to develop and validate theoretical and numerical methods that could provide reasonable prediction of accidental gas explosions inside tubes. Laboratory experiments and CFD numerical simulations of premixed hydrogen–air flames in tubes are the basis of the thesis.

In the experiments, both the dynamics of premixed hydrogen–air flame propagation and pressure build-up in half-open and closed horizontal ducts at various equivalence ratios are investigated using high-speed schlieren photography and pressure records. The high-speed schlieren device is used to record the changes both in the flame shape and position as a function of time during the combustion process. The pressure transient in the duct during the nonsteady combustion is measured using a pressure transducer. The influences of gravity, opening ratio, and equivalence ratio on the flame dynamics are also examined in the experimental investigation.

In the numerical simulations, the premixed combustion wave is simulated as a two-dimensional (2D) or three-dimensional (3D) chemically reacting flow. A dynamic thickened flame (TF) model is employed in the 2D numerical simulation to account for the premixed combustion. The chemical reaction of hydrogen and air is taken into account using a 19-step detailed chemistry scheme. The 3D numerical simulations are carried out using two numerical approaches. The first one is based on the same combustion modeling technique as that in the 2D simulation, namely the dynamic TF method. The difference in the 3D calculations is that a dynamically and locally adaptive mesh refinement is adopted, and tracks the location of the flame front. Besides, the hydrogen–air chemical reactions are taken into account using a seven-step chemistry scheme. The second one is a large eddy simulation (LES) approach together with a turbulent burning velocity model. The LES premixed combustion model is applied to gain an insight into various phenomena of flame and explain the experimental observations. The model accounts for the effects of four different physical mechanisms, i.e., flow turbulence, turbulence generated by flame front itself, diffusive-thermal instability, and transient pressure and temperature of unburned gas on the burning velocity.

The experiments show that premixed hydrogen–air flames propagating in ducts undergo more complex shape changes and exhibit more distinct characteristics compared to those of other common gaseous fuels. One of the outstanding findings is that significant distortions happen to classical tulip flame front after its full formation when equivalence ratio ranges from 0.84 to 4.22 in the closed duct. This interesting phenomenon is named as "distorted tulip flame". A distorted tulip flame is initiated as the distortions or indentations are created very near the leading tips of the tulip lips after a well-pronounced classical tulip flame is produced. The distorted tulip flame develops into a salient "triple tulip" shape as the secondary tulip cusps approach the center of the primary tulip lips and appear comparable to the primary cusp. A second distorted tulip flame appears with a cascade of secondary cusps on the primary tulip lips just before the collapse of the first one. The tulip flame distortions are specially scrutinized and distinguished from the classical tulip. The dynamics of a distorted tulip flame is different from that of a classical tulip flame. The distorted tulip flame experiences more complex shape changes and more unstable combustion process than the classical tulip flame. The normal tulip flame can be reproduced after the disappearance of the first distortion followed by another distortion. The schlieren images and the pressure records show that the distorted tulip flame propagation can be divided into five stages of dynamics, i.e., spherical flame, finger-shape flame, flame with its skirt touching the sidewalls, tulip flame, and distorted tulip flame. The initiation of flame shape changes coincides with the deceleration both of pressure rise and flame front speed for flames with tulip distortions. And the formation and dynamics of both tulip and distorted tulip flames depend on the mixture composition. Gravity has a noticeable impact on the tulip flame and can lead the tulip flame to collapse in different ways between low and high equivalence ratios. The opening ratio can significantly influence the flame dynamics in a partially open duct. When the opening ratio is smaller than 0.4, a remarkable distorted tulip flame can be formed. The characteristic times and the corresponding characteristic distances of flame front increase with the increase of the opening ratio.

The flame dynamics observed in the experiments is well reproduced in the 2D numerical simulation with the TF method. Flame-induced reverse flow and vortex motion are observed both in the experiments and the 2D simulations. The interactions between the flame front, reverse flow and vortices in the burned gas change the flame shape and ultimately drive the flame front to develop a tulip shape. The pressure wave triggered by the first contact of the flame with the side walls is responsible for the periodic deceleration of the flame front and plays an important role in the formation of the distorted tulip flame. The flame and pressure dynamics observed in the experiments are well reproduced in the 3D numerical simulations using the dynamic TF model. The predicted pressure dynamics in the numerical calculation is also in good agreement with that in the experiment. The close correspondence between the experiment and the numerical simulation demonstrates that the TF approach is quite reliable for the study of premixed hydrogen–air flame propagation in the closed duct. Both the tulip and distorted tulip flames can be created in the simulation with free-slip boundary condition at the duct walls, which implies that wall friction could be unimportant for the formation of tulip and distorted tulip flames.

The LES numerical simulations provide further understanding of the interaction between flame front, pressure wave and combustion-generated flow, especially when the flame acquires a distorted tulip shape. The dynamics of distorted tulip flame observed in the experiment is well reproduced by the LES. The numerical simulations show that large-scale vortices are generated in the burnt gas after the formation of a classical tulip flame. The vortices remain in the proximity of the flame front and modify the flow field around the flame front. As a result, the flame front in the original cusp and near the sidewalls propagates faster than that close to the center of the original tulip lips. The discrepancy in the flame propagation rate can finally lead to the formation of the "distorted tulip" flame. The LES combustion model validated previously against large-scale hydrogen–air deflagrations is successfully applied in this study to reproduce the dynamics of flame propagation and pressure build-up in the small-scale duct. It is confirmed that grid resolution has an influence, to a certain extent, on the simulated combustion dynamics after flame inversion.

On the basis of the experimental and numerical results of the interaction between flame front and pressure wave, the premixed flame dynamics for hydrogen–air mixture in the closed duct is theoretically analyzed. A theoretical model of the distorted tulip flame is suggested. The results predicted using the theoretical model are in satisfactory agreement with those in the experiments and LES. The theoretical analysis demonstrates that Taylor instability can be the substantial cause of the distorted tulip flame.

Keywords Premixed hydrogen-air flame · Flame dynamics in duct · Hydrogen safety · Tulip flame · Distorted tulip flame · Pressure wave · Vortex motion · Taylor instability

Acknowledgments

I would like to owe my sincerest gratitude to my supervisor Professor Jinhua Sun. I became his graduate student in the autumn of 2007. Professor Sun has supported me intellectually and financially through the rough road to complete this work. I am very grateful for his enthusiasm, patience, understanding, creative suggestions, and immense knowledge in combustion and flame dynamics. I also appreciate the excellent example he has provided as a successful mentor and professor.

I would further like to thank Prof. Vladimir Molkov and Dr. Dmitriy Makarov from the University of Ulster for their assistance in the LES simulations of pre-mixed hydrogen–air flame propagation. Thanks also go to my many colleagues and friends at the State Key Laboratory of Fire Science at University of Science and Technology of China for providing an excellent learning and academic environment.

My most heartfelt appreciation is reserved for my wife, Ying Xu, for her great help in the conduction and analysis of the experiments and numerical simulations, her constant encouragement, patience, thoughtfulness, and love over the years.

I would in particular like to express my sincere gratitude to my family, especially my parents, for their immense support, encouragement, and love.

It is a pleasure to thank the National Natural Science Foundation of China (NSFC). The research on the premixed flame dynamics and safety of hydrogen as an energy carrier is supported by NSFC under Grant Nos. 51406191, 51376174 and 50976110. I am also thankful for the support provided by National Basic Research Program of China (Project No. 2012CB719702).

Contents

Nomenclature

a^*_{max} Maximum non-dimensional flame surface area
A_{max} Maximum flame surface area (m^2)
CFL Courant-Friedrichs-Lewy number
c Combustion progress variable
d Distance from flame front to igniton point (m)
e Total energy (J/kg)
E Expansion coefficient
g Gravity acceleration (m/s^2)
h_m Enthalpy of species m (J/kg)
H Half of the duct widthfdfd (m)
L_{SGS} Mixing length for sub-grid scales (m)
m_0 Temperature index
n_0 Baric index
p Pressure (Pa)
p_0 Initial pressure (Pa)
Pr Prandtl number
R Flame radius (m)
R_0 Critical radius (m)
S_c Progress variable source term (kg/(m^3·s))
S_C Schmidt number
Sc_{eff} Effective Schmidt number
S_{cusp} Speed of primary tulip cusp (m/s)
S_{tip} Speed of flame leading tip (m/s)
S_t Turbulent burning velocity (m/s)
S_{L0} Laminar burning velocity at initial conditions (m/s)
S_L Laminar burning velocity (m/s)
S_L^w Wrinkled flame burning velocity (m/s)
t Time (s)
T Temperature (K)
T_0 Initial temperature (K)

u'	Sub-grid scale velocity (m/s)
u_j	Velocity component (m/s)
W	With of duct (m)
x_j	Spatial coordinate (m)
y^+	Dimensionless wall distance
Y_m	Mass fraction of species (m)
Z	Position of flame front (m)
Z_{cusp}	Position of primary tulip cusp (m)
Z_{tip}	Position of flame leading tip (m)

Greek Symbols

ε	Overall thermokinetic index
γ	Specific heat ratio of unburned gas
μ	Dynamic viscosity (Pa·s)
μ_{eff}	Effective dynamic viscosity (Pa·s)
μ_t	Turbulent viscosity (Pa·s)
ψ	LES combustion Model constant
ξ_{tip}	$= Z_{tip}/H$ reduced position of flame tip
τ	$= S_{u0}t/H$ reduced time
τ_{inv}	Reduced inversion delay time
τ_{sph}	Reduced finger shape delay time
τ_{tulip}	Reduced time of flame front inversion
τ_{wall}	Reduced time of flame front touching sidewalls
ρ	Density (kg/m^3)
ρ_b	Density of burnt gas (kg/m^3)
ρ_u	Density of unburned gas (kg/m^3)
θ	Growth rate
Ξ_K	Flame self-induced turbulence factor
Ξ_K^{max}	Maximum flame self-induced turbulence factor
Ξ_{lp}	Leading point factor
Ξ_{lp}^{max}	Maximum leading point factor

Bars

–	LES filtered quantity
~	LES mass-weighted filtered quantity

Chapter 1
Background and Introduction

1.1 Background

Combustible gases, e.g., hydrogen, methane, propane, and ethylene, are widely used in daily life and industries, especially in petrochemical engineering, power generation, metallurgy, mining, transportation, and gas supply. These gases are potentially hazardous materials in the processes of production, transportation, storage, and utilization, and have high propensity to cause fires and explosions. Generally, these accidents can result in severe losses of human life and property [1, 2]. Flammable mixture can be formed when reactive gas is accidentally released (or leaked) into air or when air enters into confined regions filled with combustible gas, such as vessels and tubes. The combustible mixture that forms prior to ignition source is usually referred to as premixed mixture (premixture). From the point of view of safety, reactive premixture is extremely dangerous since it presents a high risk of explosion. Once it is exposed to an ignition source, combustion wave will be formed, and then develops into deflagration. Under proper conditions, deflagration-to-detonation (DDT) and strong detonation can occur [1–3]. Deflagration waves move at subsonic speed, while detonations propagate at supersonic speed. The propagation speed and overpressure of a detonation wave can be as high as 2000 m/s and 20 bar, respectively [1, 2]. Thus detonations can lead to huge disasters, and consequently cause very serious damage to the surrounding facilities, structures, buildings, and humans. Basically, gas explosion is defined as a transient combustion process in an explosive mixture cloud accompanied with dramatic increase in both temperature and pressure [1]. In the meantime, plenty of expanding burnt gas is produced [4]. In contrast, the combustion process in which overpressure is negligible is called fash fire [1].

Major explosion accidents occur every year in the world [1]. For example, on April 17, 1997, a hydrogen explosion took place in a pipeline with internal diameter of 0.8 m at the production site of Hydro Agri in Porsgrunn, Norway [5]. This pipeline was used to carry CO_2 and connect an Ammonia Plant with a CO_2 plant. In

© Springer-Verlag Berlin Heidelberg 2016
H. Xiao, *Experimental and Numerical Study of Dynamics
of Premixed Hydrogen-Air Flames Propagating in Ducts*,
Springer Theses, DOI 10.1007/978-3-662-48379-4_1

the accident, an 850-m pipeline and many nearby buildings were damaged. According to the investigation results of the accident, a considerable amount of hydrogen entered the pipeline which was shut down for maintenance, although it was purged using nitrogen beforehand. The explosive mixture was ignited after six days. The pipe was broken into a number of pieces with certain distances apart. This implies that there was a detonation inside the pipeline in the accident [1, 6].

In 1970, an accident of propane leak and explosion took place in Missouri state in the U.S. [7]. At the beginning of this accident, liquid propane leaked after a pipeline for propane transportation raptured. Then the liquid propane vaporized, mixed with air and formed a huge amount of propane–air mixture. This energetic premixture flowed into a valley. Approximately 20 min later, the mixture was ignited and exploded violently. The explosion destroyed the buildings within 3.2 km. In October 1993, a huge explosion arising from ethylene leak occurred in the number one chemical plant of Beijing Yanshan Petrochemical Corporation, China, causing three dead and 200 million RMB in economic losses. In February 2000, a gas deflagration took place in a cable trench of Shandong Electric Power Corporation Ltd. Subsequently, an strong explosion in a natural gas pipeline was triggered by the deflagration, causing 15 dead and 56 injured. On November 21, 2009, an extraordinarily serious methane explosion occurred in a coal mine of Heilongjiang Longmay Minning Group Co., LTD, resulting in 108 dead, 133 injured, and 56 million RMB in direct economic losses. On November 14, 2011, a gas explosion happened subsequent to leakage of liquefied petroleum gas in a Chinese Hamburger restaurant in Xi'An of Shangxi Province, causing 10 dead and 36 injured. The accidents described above are just some examples of unwanted explosions. There have been a great number of similar accidents. Explosion accident keeps occurring and causing enormous losses to our society.

Closed vessels and tube/ducts are commonly used in industrial processes, particularly in petrochemical production. In general, combustible gases are carried and transported by closed vessels or tubes/ducts. Air may enter these confined regions and mixes with energetic gas within explosive limits due to improper operation or facility failure. Actual explosions occur in these vessels/tubes when exposed to an ignition source. According to statistics, gas explosion constitutes a high proportion in the total accidents in petrochemical, synthetic rubber, and natural gas industries, i.e., 42, 46, and 60 %, respectively [8, 9]. Moreover, the losses in human life and property in a single explosion accident are largely higher than those of other type accidents. Gas explosion in a confined region is a high-speed, transient process of reacting flow. The combustion regime may undergo transition from laminar to turbulent. Meanwhile, combustion wave can evolve from slow combustion into deflagration, and even into detonation. The flame propagation can be influenced by a variety of parameters, such as gas properties, initial temperature and pressure, geometry, boundary condition, obstacle, and combustion and hydrodynamic instabilities. Therefore, gas explosion in confined regions generally involves complex flame dynamics. Explosion of combustible gases is one of the important subjects in combustion and safety science and engineering. Gas explosion also remains to be one of the unsolved scientific problems in the research areas of combustion and safety [2, 10–16].

On the other hand, with increasing depletion of fossil fuels and serious environment pollution, the development of new energy sources has become an inevitable trend all over the world. Laboratories and industries pay more and more attentions to hydrogen as an energy carrier because of its potentially high efficiency and ultra-low harmful emissions. Nowadays, hydrogen utilization becomes increasingly widespread. For example, it was estimated that by 2040 the annual demand for hydrogen in the U.S. will reach 15 million tons [17]. There had been 150 hydrogen filling stations all over the world by the end of 2009. China also built Beijing and Shanghai hydrogen fuel filling stations as two exemplary stations. These two stations were successfully operated to refuel fuel cell vehicles during 2008 Beijing Olympic Games and 2010 Shanghai World Expo. Hydrogen can be used in fuel cells and internal combustion engines. In particular, hydrogen fuel cell vehicle is an ideal utilization of hydrogen energy system since fuel cell only produces water with extremely high energy efficiency. In addition, hydrogen has been also extensively used in current industries.

There are, however, serious challenges to overcome when using hydrogen as an energy carrier. Some unique properties of hydrogen, such as wider flammability range (4–75 % in air by volume), greater propensity to leak than other common gaseous fuels, extremely lower ignition energy (lowest ignition energy in air about 0.02 mJ), embrittlement, make it easier to cause accidental fires, explosions, or asphyxiations [18]. For example, Hindenburg explosion disaster caused a great loss at Lakehurst New Jersey in 1937 [19]. And more recently, hydrogen leak and explosion in Fukushima Nuclear Power Plant in 2011 resulted in severe damage to facilities. The safety practices in production, storage, distribution, and use of hydrogen are key issues to hydrogen energy industrialization [12, 20]. Unfortunately, safety standards which are commonly applied to the systems with small amounts of hydrogen or other hydrocarbons may not work when applied to the larger quantities of hydrogen [17, 20]. European Union carried out a pioneer research program, Hysafe (Safety of Hydrogen as an Energy Carrier), to support investigations in hydrogen safety sciences and technologies [21, 22]. Still, it is very important for the realization of hydrogen economy to conduct further research of hydrogen safety fundamentals and technologies and then develop comprehensive safety standards.

The major problems in relation to combustion are the unique characteristics of hydrogen due to its high diffusivity and reactivity. It is essential to study the combustion behaviors of hydrogen–air mixtures both for the purpose of safety and engineering applications of combustion. Compared to common hydrocarbon fuels, the combustion of hydrogen in air can behave differently, due to its unique combustion-related properties, e.g., lower ignition energy, wider flammability range, higher laminar burning velocity, higher diffusivity, smaller quenching distance, and greater extinction strain rates. For example, lean premixed hydrogen–air flame is more susceptible to a variety of hydrodynamic and combustion instabilities in comparison with common hydrocarbons.

Generally, hydrogen safety involves ignition, flame propagation, fire, and explosion in air. Therefore, premixed hydrogen–air combustion dynamics has been one of the main academic fields of numerous experimental and numerical studies. Nonetheless, the behaviors and characteristics of premixed hydrogen–air flame have not been sufficiently understood, especially in terms of the effects of various instabilities and turbulence on the combustion dynamics. The physical mechanisms underlying the interactions of the flame with these phenomena have not been clarified. There is also a lack of quantitative theory and models for the prediction of hydrogen–air flame dynamics during explosions in confined regions. In addition, premixed combustion in vessels also concerns appropriate operation of internal combustion engines and development of detonation waves for propulsion. Therefore, it is of great importance to investigate premixed hydrogen–air flame dynamics in confined regions, e.g., tubes/ducts and combustion vessels.

1.2 Accomplishments and Present Research Status

Premixed flame propagation is an important subject of combustion and explosion science and its research can date back to nineteenth century [23]. Mallard and Le Chatelier [24] conducted the pioneering works of premixed flame propagation in tubes. Although the early studies of premixed flame propagation were mainly focused on tackling various problems of explosions in coal mines, they provided a large number of qualitative results on the dynamics of premixed flames propagating in confined regions [25].

Later on, the rapid development of internal combustion engines and rocket propulsion greatly promoted the succeeding research of premixed flame propagation. This background led to booming development of many new experimental techniques and facilities, and thus made a great contribution to new discoveries across a variety of aspects of flame dynamics. Furthermore, the academic accomplishments in mass and heat transportation, chemical kinetics, and hydrodynamics greatly spurred the theoretical studies of flame propagation [16, 25, 26]. Research of combustion showed a robust growth in the early 1970s, associated with the dual concerns for energy sufficiency and air pollution. In the recent years, combustion and flame research entered a period of quantitative predictability with the emerging of laser technologies and the high-performance supercomputers [16]. With the help of laser techniques and other diagnostics (such as schlieren and shadow photography techniques), the mechanisms underlying combustion dynamics can be examined in detail, including chemical reaction kinetics, flame dynamics, and hydrodynamics. Computational fluid dynamics (CFD) has been greatly promoted by the rapid advance in computational capabilities. The applications of CFD in combustion science and engineering have been significantly facilitating the quantitative predictability in combustion and explosion.

Although significant progress has been made in combustion and explosion research and practices, many scientific issues remain unsolved. For example, new discoveries of flame propagation phenomena observed recently have not been fully understood [27, 28]. Another example is that the nonlinear development of hydrodynamically unstable flame has not been sufficiently addressed. In addition, a comprehensive understanding of interactions between flame and turbulence has not been achieved, and there is a lack of model to accurately predict turbulent combustion velocity [2, 16, 26, 29]. In terms of chemical kinetics, only relatively simple reaction mechanisms can be accurately predicted, e.g., hydrogen reaction in air or oxygen. However, it is difficult to accurately calculate the complex hydrocarbon chemical reactions, such as propane chemistry in air, especially when the background pressure and temperature vary. Furthermore, despite the rapid growth of computation power, it is still unrealistic to completely implement detailed chemistry mechanisms of large hydrocarbons and their mixtures into CFD codes [16]. Premixed flame dynamics in tubes has drawn particular attention since it prepresents the burning processes taking place in internal combustion engines, and models the flame evolution that may result in transition from deflagration to detonation in explosions [2, 12, 15, 30–33]. Moreover, mechanisms leading to the supernova explosions have much similarity to those of gas explosions [16, 33]. Therefore, premixed flame dynamics remains to be one of the research hot spots in combustion and explosion sciences because of its importance and complexity [2, 12, 15, 26, 28, 32, 34–50]. The fundamentals and accomplishments of premixed flames in confined regions that are most relevant to this work are introduced as follows.

1.2.1 Premixed Flame

Premixed flame is a very fundamental phenomenon of combustion. It can be categorized into laminar and turbulent premixed flames based on different combustion regimes. A flame is initiated and propagates into unburned mixture after ignition, forming a combustion wave. Basically, this wave can undergo an acceleration process and even develop detonation under proper conditions. Following weak ignition, premixed flames in a tube generally can evolve from laminar, wrinkled, transitional, to turbulent flames, and possibly to detonation.

1.2.1.1 Ignition of Premixed Flame

Basically, ignition of combustible mixture is a spontaneous ignition due to electric spark or contact with a hot surface. Under special conditions, explosion can be initiated by other mechanisms, for example, the ignition and subsequent explosion resulting from diffusion–ignition mechanism during high-pressure hydrogen release [51, 52]. The ignition caused by an external ignition source is generally referred to as forced ignition [53]. Most ignition methods initiate a flame by releasing sufficient energy to heat the mixture up to a temperature which is high enough to cause a spontaneous ignition. This temperature should be sufficiently high to sustain a flame after removal of ignition source. Electric spark is a common ignition approach, while other ignition techniques include hot inert gas, hot surface, laser, etc.

Ignition energy may influence the propagation of the consequent flame. Williams [54] investigated the relations between the temperature resulted from an ignition and the intensity of the ignition itself. A fast flame or detonation can be ignited by a strong ignition source. The ignition of gas mixture in industrial tubes usually arises from weak ignition sources, such as electric welding, spark, and hot tube surface. The ignition source discussed in the present work is weak ignition which initiates laminar flames. Bradley et al. [55], Lamoureux et al. [56] and Huang et al. [57, 58] studied the ignition and propagation of laminar flames of methane, natural gas, and hydrogen and demonstrated that the flame propagation becomes independent of the effects of spark ignition at a radius of about 6 mm.

1.2.1.2 Laminar Premixed Flame

Laminar flame structure and laminar burning velocity are two important fundamental parameters of premixed flame propagation. In most cases, chemical reactions occur in local laminar flame sheet even in turbulent combustion. The effects of turbulence generally result from the disturbance and wrinkling of turbulent flow imposed on a laminar flame. Early in 1960s, Fristrom and Westenberg [59] presented a detailed experimental study on laminar flame structure. Laminar flame structure and burning velocity are mainly determined by the chemical kinetics, thermophysical properties, and transport properties of species inside a flame front. Most of the recent studies used numerical approaches to calculate laminar flame structure and laminar flame speed. These methods solved the governing equations which consist of conservation equations of mass, momentum, energy, and species [60]. The thermophysical and transport properties can commonly be included in the database of numerical solvers.

(1) Structure of laminar premixed flame

Structure of a one-dimensional (1D) steady-state laminar premixed flame is schematically shown in Fig. 1.1. The flame can be divided into four zones from left to right as follows [61]:

Unburned mixture zone. This zone comprises cool unburned fuel–oxidizer mixture. The profile of each variable is flat.

Preheat zone. The heat released due to chemical reactions is conducted toward the low-temperature mixture next to the reaction zone. The preheating process has a positive effect on the chemical reaction of the unburned mixture in this zone.

Reaction zone. Chemical reactions of combustion take place in this zone. The combustible mixture is consumed in the region, while combustion products and heat are generated. The heat released from the reactions is mainly taken up by the expanding combustion products. For subsonic combustion in an open space, the pressure difference across a premixed flame is very small since it can be relieved

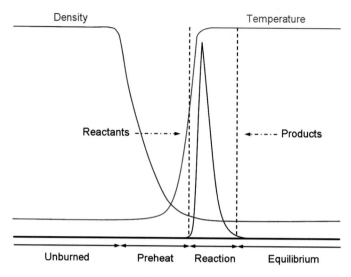

Fig. 1.1 Sketch of laminar premixed flame structure

easily. The density of burnt matter is largely lower than that of unburned mixture due to expansion effect.

Equilibrium zone. This zone is a burnt region with all variables in an (relatively) equilibrium state. Similar to the unburned mixture zone, the profiles of variables are flat.

(2) Thickness of laminar premixed flame

The flame thickness consists of reaction zone and preheat zone. Following the similarity principle, laminar premixed flame thickness δ_L can be obtained as follows [29, 37]:

$$\delta_L = \frac{\lambda_u}{\rho_u c_p S_L},$$ (1.1)

where S_L, λ_u, ρ_u, and c_p are the laminar burning velocity, thermal conductivity, density, and specific heat at constant pressure of unburned mixture, respectively. The flame thickness given by this equation is usually referred to as diffusion thickness. It is much smaller than the thickness of a realistic flame [29]. A more practical way to calculate laminar premixed flame thickness can be based on temperature profile as follows:

$$\delta_L = \frac{T_2 - T_1}{\max(|\partial T / \partial x|)},$$ (1.2)

where T_2 and T_1 are the highest temperatures in burned and unburned regions, respectively. The term $\max(|\partial T/\partial x|)$ denotes the absolute value of the maximum temperature gradient.

If the thermal conductivity is calculated using Sutherland's law, Eq. (1.2) can be simplified as [29]:

$$\delta_L = \frac{2\lambda_u}{\rho_u c_p S_L}\left(\frac{T_2}{T_1}\right)^{0.7}. \tag{1.3}$$

A quite reasonable flame thickness can be obtained by Eq. (1.3) instead of numerical computation when the temperature of the flame is known.

Laminar flame thickness is very small, generally a fraction of 1 mm [29, 62]. For example, the flame thicknesses of hydrogen–air mixtures at equivalence ratios 0.7 and 4.0 at normal temperature and pressure are 0.35 mm and 0.43 mm, respectively [62]. The parameters influencing flame thickness include mixture composition, temperature and pressure of mixture, etc.

(3) Laminar burning velocity of premixed flame

Laminar burning velocity is a crucial parameter in combustion and flame dynamics. Particularly, it is very important in the theoretical analysis and numerical simulation of flame dynamics. With the help of dimensional analysis, Landau and Lifshitz [63] constructed a relationship between laminar burning velocity, thermal diffusivity, and chemical reaction time as

$$S_L \propto \sqrt{\frac{\lambda_u}{\rho_u c_p \tau}}. \tag{1.4}$$

Following this relationship, Bychkov and Liberman [37] suggested a formula for calculating laminar burring velocity:

$$S_L = \sqrt{\frac{2\lambda_u}{\rho_u c_p \tau}\frac{\sqrt{E}}{E-1}\frac{T_b}{E}}\exp(-\frac{E}{2T_b}), \tag{1.5}$$

where E is the expansion coefficient, namely the ratio of unburned density to burnt density, and T_b is the temperature of burnt gas.

There are a large number of experimental and numerical studies of laminar burning velocity of combustible mixtures [40, 62, 64–69]. Dahoe [69] gave the laminar burning velocity of hydrogen–air mixture as a function of equivalence ratio (composition) based on various references [62, 64, 70–79], as shown in Fig. 1.2.

The temperature and pressure of unburned mixture have significant effects on flame structure and burning velocity [29, 80]. For the flame propagation in confined regions, the pressure and temperature of unburned mixture increase as flame propagates due to the compression of combustion wave. The dependence of laminar

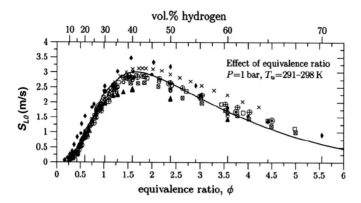

Fig. 1.2 Laminar burning velocity of hydrogen–air mixture as a function of equivalence ratio/hydrogen concentration by volume. Reprinted from Ref. [69], Copyright 2005, with permission from Elsevier

burning velocity on temperature and pressure during flame propagation can be taken into account based on an assumption of adiabatic compression [29]:

$$\frac{S_L}{S_{L0}} = \left(\frac{T}{T_0}\right)^m \cdot \left(\frac{p}{p_0}\right)^n, \tag{1.6}$$

where T and p are the temperature and pressure of unburned mixture after compression, and T_0, p_0, and S_{L0} are the temperature, pressure, and laminar burning velocity under initial conditions, respectively. And m and n are the temperature and baric indices independent of the burning velocity, respectively. Flame propagation in confined region is usually fast and thus the propagation time in can be rather short. If the combustion process is assumed to be adiabatic, temperature can be related to pressure according to the adiabatic compression law as follows:

$$\frac{T}{T_0} = \left(\frac{p}{p_0}\right)^{(\gamma-1)/\gamma}, \tag{1.7}$$

where $\gamma = c_p/c_v$ is the specific heat ratio, c_p and c_v are the specific heats at constant pressure and constant volume, respectively. During adiabatic compression of unburned matter, pressure and temperature change simultaneously. Therefore, the laminar burning velocity can be defined as a function of pressure alone [81]:

$$S_L = S_{L0}\left(\frac{p}{p_0}\right)^\varepsilon, \tag{1.8}$$

where ε is the overall thermokinetic index.

1.2.1.3   Flame Instabilities

Flame instabilities include hydrodynamic instability, thermal-diffusive instability, Rayleigh–Taylor instability, Kelvin–Helmoltz instability, and gravitational instability. Flame instability is one of the hot spots of combustion research since it is essential for flame dynamics [26, 40, 82, 83]. Current quantitative theory of flame instability is mainly based on linear analysis, whereas nonlinear evolution of flame instability has not been sufficiently understood [26].

(1) Hydrodynamic instability

Hydrodynamic instability was proposed by Darrieus [84] in 1938 and Landau and Lifshitz [85] in 1944 separately. Thus it is also called Darrieus–Landau (DL) instability. This instability results from gas thermal expansion during combustion.

It is assumed in the DL theory that the flame front is infinitesimally thin and moves normal to itself at a constant velocity relative to the unburned mixture. The flame is not affected by hydrodynamic disturbances. Within the above assumptions, the growth rate of DL instability can be yielded by the analysis of linear stability of a planar flame [26]

$$\omega_{LD} = \frac{-E + \sqrt{E^3 + E^2 - E}}{E + 1}. \tag{1.9}$$

The physical origin is shown in Fig. 1.3. If a planar flame is slightly perturbed, the streamlines in the burnt gas diverge behind the concave part of the flame front and converge behind the convex part. The jump in transverse velocity implies that the curved flame is equivalent to a flat sheet of vortices. The deflection of streamlines leads to a pressure gradient that creates an additional displacement of the flame front, and consequently increases the initial curvature [26, 30].

Following the DL linear stability analysis, a planar laminar front is intrinsically unstable due to the hydrodynamic instability for all wavelengths of disturbances [26, 30]. However, this conclusion is not valid for the perturbations of short wavelengths which are comparable to the flame thickness, since they can cause

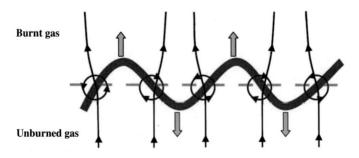

Fig. 1.3 Schematic graph of the physical origin of hydrodynamic instability. Reprinted from Ref. [26], Copyright 2009, with permission from Elsevier

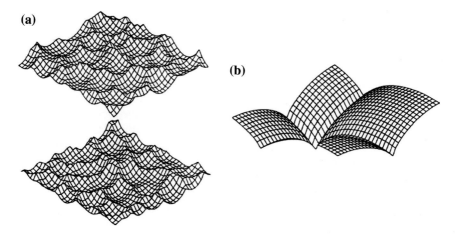

Fig. 1.4 Numerical results of cellular flame structures arising from **a** thermal-diffusive instability, and **b** hydrodynamic (DL) instability [40]. (Law CK. Combustion physics. Cambridge University Press, 2006, p. 466)

distortions to the flame front. These effects were not taken into account in the DL theory. In addition, the DL instability due to thermal expansion may be stabilized by diffusion processes within flame front. DL instability is always present during premixed propagation in a confined space with a characteristic length scale significantly larger than the flame thickness. The DL destabilizing effect leads to cracked surface of an expanding flame and continuous wrinkling of flame front. It has been found by numerical computations that DL instability creates steady cells of regular shapes and sizes [40], as illustrated in Fig. 1.4b.

(2) Thermal-diffusive instability

Thermal-diffusive instability is a consequence of the difference between the thermal conductivity of mixture and the molecular diffusivity of the deficient reactant. For example, lean hydrogen–air mixtures are very susceptible to this instability due to the high molecular diffusivity of hydrogen [86]. The thermal-diffusive instability can cause local changes in mixture composition and reaction rate, and subsequently flame front wrinkling. The wrinkling effects eventually contribute to augmentation of burning velocity [26, 40]. The Lewis number is usually used to describe the difference between molecular transport coefficients:

$$Le = \frac{a}{D_d}, \qquad (1.10)$$

where a is the thermal conductivity of mixture and D_d is the molecular diffusivity of the deficient reactant.

For mixture with $Le < 1$, flame is perturbed by the thermal-diffusive instabilities and wrinkles grow over the flame surface. Thermal-diffusive instabilities can also enhance DL instability. These effects lead to increase in flame surface area and consequently burning velocity. It has been put into evidence that thermal-diffusive instability generates cells of chaotic nature on flame front, as shown in Fig. 1.4a. This flame structure may result in turbulence in a flame, which is referred to self-turbulization of flame [87]. If $Le > 1$, the thermal-diffusive effects may lead to stabilization of a flame against the DL instability. The flame surface can thus stay smooth, with the burning velocity close to the initial laminar burning velocity.

(3) Flame instabilities induced by confinement and obstructions

Pressure waves (or acoustic waves) can be generated during flame propagation in confined regions. These waves are reflected when they reach the obstacles and/or walls, and travel forth and back inside the region. The interactions of the pressure waves with the flame front can lead to development of flame disturbances and instabilities [88]. The acoustic instabilities can be eliminated by lining the walls of confined region with special materials that can absorb pressure waves [89].

The acoustic effects may enhance or suppress flame wrinkling. The acoustic flame instabilities are related to the baroclinic term in the vorticity equation, $\nabla p \times \nabla \rho / \rho^2$, where ∇p and $\nabla \rho$ are the pressure gradient and density gradient, respectively. When the local angle between ∇p and $\nabla \rho$ is larger than 90°, vorticity generation becomes stronger. On the contrary, when the local angle between ∇p and $\nabla \rho$ is less than 90° and not equal to 0°, the vorticity generation reduces. In other words, the flame wrinkling is enhanced if pressure gradient is adverse since vorticity increases, whereas the flame wrinkling is suppressed when the pressure gradient is positive. Generally, varying pressure field results in intense generation of vorticity (via baroclinic effects), which then leads to enhancement of flame wrinkling. This was corroborated by numerical simulations [90]. In addition, the study by Bradley and Harper [91] showed that the Rayleigh–Taylor (RT) instability arising from the baroclinic effects can be favorable for the generation of turbulence.

RT instability and Kelvin–Helmoltz (KH) instability can play an important role in the flame dynamics in confined or obstructed reactions. The RT instability is usually driven by forces associated with the different gas densities on the two sides of the flame front. There is a strong misalignment of the pressure gradient and density gradient in a perturbed flame when exposed to an acoustic wave. This misalignment creates a baroclinic torque, which generates vorticity. This physical mechanism is also known as Richtmyer–Meshkov (RM) instability in high-speed compressible flows. KH instability occurs when there is a velocity shear in the flow field. Both RT and KH instabilities can be initiated when a flame is accelerating over an obstacle or through a vent [30].

1.2.1.4 Turbulent Premixed Flame

In explosions, a laminar flame, which forms after weak ignition, would be perturbed by various hydrodynamic and combustion instabilities as mentioned above. The wrinkled flame then develops into cellular flame that can finally cause turbulent combustion [30, 86, 88]. In addition, turbulent flame can also result from interactions of flame front with walls, obstacles, and turbulence of incoming flow. For the flames propagating in smooth channels (without obstruction), if the background flow is initially quiescent, turbulence would be principally induced by flame instabilities and boundary layer effects [2, 15, 92]. One of the main effects of turbulence on premixed combustion is that turbulence can largely increase burning velocity. The increase of flame speed due to turbulence is closely connected to the intensity of turbulence or turbulent fluctuation velocity [29].

The interaction between flame and flow turbulence mainly depends on the length and time scales of both the flame and turbulence. The length scale of a flame can be characterized by laminar flame thickness δ_L, while the time scale can be defined as the ratio of laminar flame thickness to laminar burning velocity $\tau_L = \delta_L/S_L$. It is known that the length scale and time scale of turbulence are distributed continuously in the flow field. Turbulent flows can be viewed as an entire hierarchy of eddies over a wide range of length scales and time scales. The length scale and axis location of these eddies are stochastic [93]. The integral scale of turbulent flow l_0 can be defined as the length scale of the largest eddy. This scale is commonly smaller than the characteristic scale of the physical domain considered, such as the diameter of a tube. The corresponding time scale is τ_0. Large eddies break into smaller scale structures. This process continues, producing smaller and smaller eddies, which creates a hierarchy of eddy structures, until molecular diffusion becomes important so that viscous dissipation of energy eventually takes place. The length scale at which the viscous dissipation occurs can be taken as the smallest scale of turbulent flow. It is usually referred to the Kolmogorov length scale l_k [94], and the corresponding time scale is the Kolmogorov time scale τ_k.

The effects of turbulence on premixed combustion can be described using a diagram of combustion regime that is commonly called Borghi diagram [95]. Figure 1.5 is an example of Borghi [96] which is a log–log diagram. The vertical axis is the dimensionless velocity u'/S_L, which is derived by dividing turbulent fluctuation velocity u' by laminar burning velocity S_L. The horizontal axis denotes the dimensionless length scale l_0/S_L, which is the ratio of turbulent integral scale to laminar flame thickness. Combustion regime is usually illustrated using Borghi diagram since turbulence spectra is very wide. Other types of Borghi diagrams can be found elsewhere [30, 37, 97].

There are five combustion zones in Fig. 1.5. The line $R_T = 1$ divides combustion into laminar and turbulent combustion zones. R_T is the turbulent Reynolds number defined as $R_T = [(u'l_0\rho)/\mu]^2$ [96], where μ is the kinematic viscosity. In the zone with $u' < S_L$, turbulent fluctuation velocity is smaller than laminar burning velocity, so that turbulence develops slower than flame and can only wrinkle the flame front. The two lines $K_a = 1$ and $D_a = 1$ are used to divide the turbulent combustion zones, where $K_a = \tau_L/\tau_k$ is the turbulence Karlovitz number and $D_a = \tau_0/\tau_L$ is the turbulence Damköhler number, respectively. If $K_a < 1$ and $u' > S_L$, flame thickness is smaller than the Kolmogorov length scale of eddy and the turbulence intensity is larger than laminar flame speed. The wrinkles of flame caused by this type of turbulence interact with each other, resulting in pockets or islands of burnt and

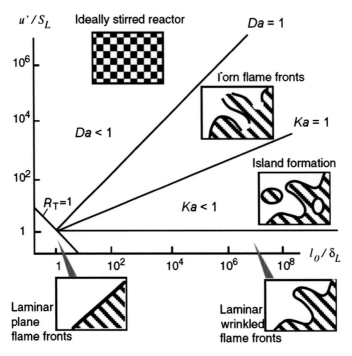

Fig. 1.5 Borghi diagram of combustion regimes. Reprinted from Ref. [96], with kind permission from Springer Science + Business Media

unburned gases. Peter [97] referred this zone as wrinkled flamelet zone. If $K_a > 1$ and $D_a < 1$, Kolmogorov eddy length scale is smaller than flame thickness. In this case, laminar flame cannot be sustained because the smallest eddies change the internal structure of flame. Local quenching phenomenon of flame occurs when Kolmogorov eddies stretch the flame intensely. This zone is also known as torn flame front zone or distributed reaction zone. In the last zone with $D_a > 1$, chemical reaction is slower compared to the turbulent velocity. Even larger eddies can enter into the reaction zone with higher speed than laminar burning velocity. This situation is the ideally stirred combustion.

In addition to the effects of turbulence wrinkling, flame stretch can also change the flame surface area, leading to variation in local laminar burning velocity. In general, flame stretch rate can be defined as [40]

$$K = (\frac{1}{A_f}) \frac{dA_f}{dt}, \tag{1.11}$$

where A_f is the flame surface area.

When flame front interacts with the surrounding vortex motion, the following two aspects need to be considered to yield the flame stretch rate: (1) differential of the tangential velocity of flame surface which gives a stretch rate equal to the aerodynamic strain rate; and (2) stretch of flame itself caused by the mixing effects during the propagation of wrinkled flame front toward unburned mixture. This stretch is known as curvature term. Law et al. [40] suggested a model of flame stretch rate by combining these two physical phenomena:

$$K_{2D} = \frac{\partial u_T}{\partial s} + \frac{S_L}{R}, \qquad (1.12)$$

where R is the local curvature radius of flame. The first right term of Eq. (1.12) is the stretch effects due to aerodynamic strain, and the second term on the right side represents the curvature term.

1.2.1.5 Detonation

Detonation is a combustion phenomenon very different from deflagration. Deflagration is a combustion process with subsonic flame propagation speed. The combustion can be laminar or turbulent. Deflagration is dominated by thermal conduction, molecular transport, and turbulent transport. In contrast, detonation is a combustion process with reaction wave propagating at supersonic speed relative to the unburned mixture ahead. A detonation wave is composed of a precursor shock wave and a chemical reaction zone that follows immediately behind the shock. The shock wave compresses the unburned gas in front of it and quickly increases the temperature of the gas. The compressed unburned mixture ignites spontaneously after an induction time. A large amount of heat is released from the exothermic chemical reaction in the reaction zone. This leads to rapid gas expansion which in turn sustains the precursor shock wave. It indicates that the precursor shock wave couples with the chemical reaction zone. Therefore, detonation is a self-sustained process.

The mathematical calculation of detonation velocity can be given by the classical Chapman–Jouguet (CJ) model [98, 99]. In the model, detonation is treated as a strong 1D discontinuity with chemical reaction. The detonation wave is assumed to be a control volume in which the chemical reaction zone is immediately adjacent to the shock wave. It also assumes that the chemical reaction rate is infinite and the speed of the burnt gas immediately behind the detonation wave is sonic. The detonation velocity is obtained by solving the 1D conservative equations. Furthermore, the CJ theory relates the gas state of combustion to the tangent intersections between Rayleigh line and Hugoniot curve. Although CJ model does not account for chemical kinetics, the detonation velocity given by this model is in good agreement with experimental measurement [53]. Another classical detonation model is the ZND model which was separately proposed by Zeldovich [100], von Neumann [101], and Döring [102] (ZND). Similar to the CJ theory, the ZND theory also assumes that precursor shock wave leads detonation with a CJ theoretical velocity. What is different is that the ZND model takes into account finite chemical reaction rate. Thus the induction distance between the precursor shock and chemical reaction zone can be obtained. Both the CJ and ZND theories assume detonation as a 1D plane. However, experimental observations have revealed that detonation front is a complex and unstable 3D structure [30, 63, 103, 104]. The instabilities of detonation wave arise from secondary shock structures which travel transversely relative to the detonation front. These transverse shock waves collide with each other periodically and cause periodic oscillations to the detonation front. The intersection between transverse shock and detonation wave is known as triple point. The thickness of a detonation wave is quite large compared to a common flame [3].

In the explosions in confined regions such as tubes, detonation usually takes place after the flame undergoes fast acceleration and DDT. DDT is a more complex process and has given rise to many investigations. Han et al. [105] carried out a series of experiments of gas explosions focusing on the characteristic distances of DDT. It was shown that the initial pressure and dilution gas have considerable impact on the initiation distance of DDT. Ciccarelli and Dorofeev [30] presented a review of studies on flame acceleration, DDT, and detonation in both smooth and obstructed ducts. The emphasis of the review was put on experiments of gas-phase flame acceleration and DDT. Xia et al. [106] experimentally studied the influence of a 90° curved section on the detonation propagation in a cylindrical tube. The results suggested that the curved structure of tube can increase the detonation velocity and pressure. Yu et al. [107] conducted an experimental investigation of the effects of obstacles and equivalence ratio on premixed flame acceleration and DDT in hydrogen–air mixtures in a half-open duct. The experiments indicated that equivalence ratio has an important influence on flame acceleration and DDT. When equivalence ratio is larger than 0.34, the flame accelerates fast. DDT can occur at a proper equivalence ratio.

Gas-phase detonation has been studied for many years [30]. Experimental and numerical works have presented a large number of conclusions and findings, such as detonation limits, detonation velocity, the effects of fuel types, initial conditions and boundary conditions, etc. Detonation theories revealed the relationships between detonation, hydrodynamics, chemical kinetics, and thermal physics. Nevertheless, there is a common weak point in all the current theoretical models. That is, the sub-model of chemical kinetics is ambiguous. The lack of reliable chemical models is one of the key issues that barriers the development of detonation theories since the chemical induction time is an essential parameter in chemical models and is closely related to chemical kinetics.

1.2.2   Research of Dynamics and Mechanisms of Premixed Flame Propagation in Tubes

1.2.2.1   Research of Premixed Flame Dynamics in Tubes

Premixed flame propagation in tubes has been studied for more than one hundred years [25, 108]. Mallard and Le Chatelier [108] performed the first experimental investigations of premixed flame propagation during gas explosion in pipes in 1883. In their experiments, premixed gas mixture was ignited at the closed end of a half-open pipe and then a premixed flame propagated toward the open end. They found that a series of flame front inversions occurred along the pipe axis during the propagation process. The first shape photographs of a premixed flame front propagating irregularly in tubes were published by Ellis in 1928 [109]. He observed that the shape of the flame changes suddenly from a forward pointing finger to a backward pointing cusp for closed tubes. This curious phenomenon was named as "tulip flame" in 1959 by Salamandra et al. [110].

There have been a large number of studies of tulip formation and its possible mechanisms [37, 42, 43, 45, 48–50, 110–114]. Nevertheless, the mechanism of tulip formation has not been conclusively determined. Dunn-Rankin and Sawer [50] thought, based on experiments, that tulip flame is a combination of a series flame propagation processes. Clanet and Searby [114] divided the tulip flame propagation into four stages in their experimental work and proposed an empirical model for the tulip flame evolution that can estimate the characteristic time and position of flame tip of each stage. Following this study, Bychkov et al. [36] suggested a theoretical model for the early flame acceleration and tulip flame formation. They also pointed out that the formation of tulip flame does not depend on Reynolds number. The experiments by Ellis [109] also show that the flame experiences several stages in the early phase, and the flame shapes at these stages are influenced by the length of tube. Schmidt et al. [115] found that propane–air flame propagates in an oscillating way in tube using schlieren technique. Starke and Roth [45] investigated the flame propagation and oscillating behavior in ethylene–air mixtures in closed tubes using photography and laser Doppler anemometer. Flame–acoustics interaction is also important for the flame dynamics. Gonzalez [116] numerically demonstrated that the periodically oscillating behavior of a premixed flame propagating in a closed tube at the later stage results from the interactions between flame and acoustic waves. And the periodic flame acceleration/deceleration drives the flame to display an oscillating cellular pattern, which is akin to Taylor instability. In fact, flame dynamics in tubes is sensitive to various parameters, such as equivalence ratio, expansion coefficient, burning velocity, tube geometry, boundary conditions, and initial conditions [45, 50, 111, 112, 116, 117]. Markstein [118] carried out experiments of the interactions between shock waves and laminar premixed flames. In his experiments, flame front inverted after a shock wave passed through, creating a tulip-like flame shape. Markstein [118] suggested that the direct reason for the flame inversion is the sudden deceleration of flame front caused by the shock wave. He explained this with the help of Taylor instability.

With respect to the instabilities of flame propagation, Barrere and Williams [119] thought that the mechanisms of flame instabilities can be categorized into internal and external instabilities. Guenoche [120] performed experimental studies of flame propagation in tubes, and examined a variety of factors which may influence the flame propagation, flame instabilities, and the interactions of flame front with acoustic waves. Chen et al. [121] studied the effects of rarefaction waves on premixed propane–air flame propagation in a rectangular duct using high-speed schlieren cinematography and found that rarefaction wave can induce more insta-bilities. Ye et al. [122] investigated interaction between a spherical flame and shock wave using shadowgraph technique. The experimental results showed that the flame developed Richtmyer–Meshkov instabilities due to the effects of the shock waves. Fan et al. [123] pointed out that interaction of flame with shock wave plays an important role in turbulence generation, flame acceleration, and DDT. Besides, the coupling between flame and flow has also attracted attentions since the interaction between a flame front and the flow induced by itself has a significant effect on the flame shape and stability in a tube [43, 49, 50, 111, 113, 117].

There have been also numerous studies on effects of obstacles on flame prop-agation in tubes since obstruction can greatly influence the flame acceleration and turbulization [13, 14, 30, 46, 124–128]. Fairweather et al. [13] experimentally and theoretically investigated the effects of obstacles on premixed methane–air flame propagation in moderate-scale ducts. They found that the flames propagating in smooth ducts were laminar, whereas the flames in obstructed ducts underwent transition from laminar to turbulent. Moreover, the blockage ratio had an important influence on the flame dynamics. The shear flow and vortex motion induced by obstacles were the causes of turbulent flames. Dorofeev et al. [32, 88] and Lee et al. [129] studied turbulent flame propagation in obstructed tubes. They suggested three different turbulent combustion regimes: quenching regime, slow combustion regime (slow turbulent flame), and chocking regime (fast turbulent flame). The studies by Dorofeev et al. [32] and Kuznetsov et al. [130] show that the final turbulent combustion regime mainly depends on the expansion ratio of the mixture consid-ered. They thought that expansion ratio plays a key role in the formation of strong acceleration and fast turbulent flame. Masri [131] and Ibrahim et al. [132] conducted a parametric experimental study of the effects of obstacle geometry (shape and size), vent features and blockage ratio on gas deflagration, and overpressure in tubes. The results indicate that the overpressure has a positive correlation with blockage ratio and venting pressure, while the growth rate of overpressure is closely connected to the obstacle geometry. The overpressure growth rate with rectangle obstacles is higher than that with cylindrical obstacles. Patel et al. [133] experi-mentally and numerically examined premixed flame propagation in a half-open vessel with obstacles. They found that the stretch effects could be dominant in flame turbulization and acceleration between serial obstacles. Johansen and Ciccarelli [134] investigated influence of blockage ratio on both the early flame acceleration and the flame front dynamics along the axis of combustion chamber using a novel schlieren technique. It was shown that the higher the blockage ratio is, the larger the flame acceleration is. Zhou and Li [135] experimentally detailed the effects of ignition energy, tube size, and obstacle on premixed propane–air flame propagation in a duct. They suggested that the ignition energy only has an influence on the initial flame propagation. The flame acceleration can be promoted by increasing tube size. Lin et al. [136] and He et al. [137] studied effects of obstructions on the premixed flame propagation during methane explosions. They concluded that the main effect of obstruction on flame propagation is the positive feedback of turbulent flow induced by obstacles to burning velocity.

Under proper conditions, a premixed flame propagating in a tube may undergo DDT [2, 14, 30, 138]. The study of DDT phenomenon has started since the end of nineteenth century [108]. Urtiew and Oppenheim [30] studied DDT using strobo-scopic schlieren technique in 1966 and they also called detonation as explosion in explosion. Zeldovich et al. [139] proposed a theory to describe the detonation initiation in unburned gas and took ignition delay gradient as the determining factor of DDT in 1970. Lee and Dorofeev [30] suggested a similar theory in 1978. This theory was based on the shock waves generated by gradient of chemical induction time and shock wave amplification by coherent energy release (SWACER). On the basis of numerous experiments, some empirical criteria were suggested [15, 30, 88]. Most of these criteria are related to geometric size of the region considered and the reactivity of mixture. Furthermore, Dorofeev [140] thought that supersonic com-bustion may be triggered before DDT in a closed region with many obstacles. Gamezo et al. [128] suggested that the flame–shock interaction, flame–vortex interaction, RT instability, RM instability, and KH instability are the dominant factors for DDT initiation in obstructed channels.

Geometry of tube may assume an important role in flame/detonation propagation and it is worth noting some of the studies where curved tubes were used. Zhou et al. [141] investigated the dynamics of premixed flames propagating in a closed duct with a 90° bend using high-speed photography and 3D numerical simulations (based on KIVA-3 V). Their results indicate that the flame shape changes in the bend are mainly due to vortex motion near the inner wall and the high-pressure effect at the outer wall. The flame shows a shedding behavior when entering the first 45° of the curved section. The combustion-generated flow becomes more compli-cated due to the penetration effect of flame into the unburned mixture in the bend. This flow in turn interacts with the flame. Sato et al. [142] conducted a study of propane–air flame propagation in a small-scale open square duct with a 90° curved section using high-speed schlieren photometry and 2D numerical simulation based on incompressible model without chemical reaction. It was found that the flame is distorted as it propagates in the curved section and the effects of bend on the flame behaviors are attributed to the flow conditions of the unburned gas ahead of flame front. They concluded that the flame dynamics around the bend is primarily determined by the flow nature of the unburned mixture. The flow pattern can be correctly reproduced in the 2D numerical simulation in the absence of secondary flow. Wang et al. [143–145] performed an experimental study of gas-phase deto-nation waves propagating in semi-circular curved and T-shaped tubes. They found that the compression waves and rarefaction waves in a curved section conspire to distort the detonation, causing thinner reaction zone and higher shock intensity near the inner wall than those near the outer wall. It was also shown that the cellular detonation structure disappears first and then reappears around the T-shaped sec-tion. The structure of detonation wave changes irregularly and there is a second ignition when the detonation moves through the T-shaped section. He [146] studied propane–air flame propagation in a closed duct with a 90° bend by high-speed schlieren photography. The effects of various parameters, such as equivalence ratio and ignition location, on the flame propagation were analyzed. The flame shape, flame temperature, flame propagation speed, and pressure build-up were given.

1.2.2.2   Combustion Regimes During Flame Propagation in Tubes

In general, flames following weak ignition may undergo various propagation/combustion regimes [140], as shown in Fig. 1.6. Five different regimes or stages may be distinguished: (1) weak ignition, (2) laminar flame, (3) wrinkled laminar flame, (4) turbulent flame, and (5) quasi-detonation/detonation. The flame first develops a smooth laminar flame after a combustible mixture is ignited by a weak ignition. The flame in this regime is controlled by laminar burning velocity and expansion ratio. As the flame continues to propagate, the flame surface area increases quickly. A flame is actually a discontinuity of density and is susceptible to intrinsic instabilities [26, 84, 85]. The flame can develop wrinkled and cellular structures due to the disturbances of flame instabilities mentioned above. The wrinkling effects lead to increase in flame surface area and thus accelerate the flame. The flame acceleration would be further promoted as turbulence occurs and grows in the flow. Turbulence can be generated by several mechanisms, such as wall effects, flame instabilities, and interaction of flame with pressure waves. The wall effects include shear flow, rough

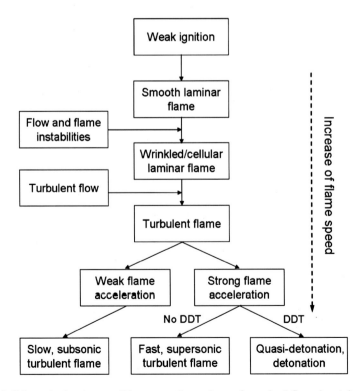

Fig. 1.6 Schematic showing possible propagation regimes of premixed flame in a tube [140]

boundary effects, and turbulization due to irregular shape of wall. Flame instabilities may lead to self-turbulization of flame front. In closed tubes or obstructed regions, turbulence can also be generated by the interactions between flame and acoustic waves. The variations in tube geometry, such as diameter and curvature, may induce more turbulence. The flame surface and heat and mass transfer rates are increased due to turbulent wrinkling and transport. The overall burning rate is consequently increased. In turn, more turbulence is induced as the flame speed increases. Therefore, the interaction between turbulent flow and flame is a process of positive feedback. Nonetheless, the flame cannot accelerate infinitely because local stretch and curvature effects of flame can reduce the local burning velocity. Moreover, flame can be quenched by these effects in the presence of flow with high turbulent intensities. The reason for this is that burning velocity and flame temperature are decreased by large stretch and fast mixing of hot gas with cold gas. On the other hand, flame may decelerate when it starts to reach tube walls since flame surface area can be reduced considerably by the quenching effects of wall.

DDT may occur when the flame propagation develops a deflagration mode. Generally, DDT takes place in the flow field behind shock waves which are ahead of flame front. Compression waves are generated by an accelerating flame. The compression effects of these pressure waves results in temperature increase in unburned mixture. The ignition delay time reduces due to the increase of temperature. Temperature distribution in a turbulent flow is usually non-homogenous. The unburned regions with high temperature have small ignition delay time. Spontaneous ignition occurs first in these regions. Consequently, new flames are initiated and more pressure waves are generated. These new pressure waves further increase the temperature of unburned gas. The leading pressure wave is strengthened when these pressure waves overtake and merge with it. These processes repeat until the leading pressure wave develops into a shock wave that is strong enough to trigger DDT and maintain a detonation.

The research of combustion and explosion safety in industries is mainly focused on turbulent combustion and DDT processes. These two combustion regimes are extremely complex and remain as tough problems worldwide and as hot spots of research [2, 14, 30, 82, 140]. Laminar combustion and flame propagation have received relatively less attention because they contribute a low proportion in industrial combustion. Nevertheless, laminar flame commonly represents the early flame acceleration in tubes and thus plays an important role in flame dynamics. Furthermore, laminar flame is a fundamental subject of combustion and explosion science.

1.2.3   Research of Combustion and Explosion Safety in Utilization of Hydrogen Energy

Hydrogen as an energy carrier is a promising alternative fuel. Hydrogen has properties which can be either favorable or unfavorable for safety compared to common hydrocarbon gas fuels. The properties favorable for safety are high diffusivity, large buoyancy, and low energy per unit volume. There are more properties unfavorable for safety, including extremely low ignition energy, very wide range of flammability limit, great propensity to leak, embrittlement, high burning velocity, and high energy per unit mass.

1.2.3.1 Hydrogen Leak

Hydrogen is the smallest element in the universe and has a very low density at normal temperature and pressure, so that hydrogen has a higher propensity to leak than other gases and liquids. In addition, the leak rate of hydrogen is also higher than hydrocarbons such as natural gas. Therefore, it is easy for hydrogen to leak from the fissures or cracks of fuel pipe, valve, and high-pressure storage tank. Table 1.1 shows the leak rates of hydrogen in comparison with propane and natural gas [147].

It can be seen from the table that the leak rate of hydrogen is very high under laminar condition, about 26 % higher than natural gas. And the hydrogen leak rate is much higher under turbulent condition, approximately 2.8 times higher than that of natural gas. The velocity of both hydrogen and natural gas can be sonic when leaking from a pressurized tank. Again, the velocity of hydrogen is about 1308 m/s, much larger than that of natural gas, 449 m/s [18]. Therefore, hydrogen, particularly high-pressure hydrogen, has a great propensity to leak and very large leak rate.

1.2.3.2 Hydrogen Properties Related to Combustion and Explosion

The flammability range of hydrogen in air is 4–75 % by volume. The lowest ignition emery of hydrogen–air mixtures is 0.019 mJ. Furthermore, the laminar burning velocity of hydrogen is much larger than that of common hydrocarbon gases. Table 1.2 gives the comparison of combustion properties between hydrogen, natural gas, propane, and gasoline [18, 19]. It shows that hydrogen has the widest

Table 1.1 Leak rates of hydrogen and propane relative to natural gas [18]

	Natural gas	Hydrogen	Propane
Parameters			
Diffusivity in air ($cm^2 s^{-1}$)	0.16	0.61	0.10
Viscosity at 0 °C (Pa s $\times 10^{-7}$)	110	87.5	79.5
Density at 70 °F, 1 atm (kg m^{-3})	0.666	0.08342	1.858
Relative leak rate			
Diffusive	1.0	3.8	0.63
Laminar	1.0	1.26	1.38
Turbulent	1.0	2.83	0.6

Table 1.2 Comparisons of combustion properties between hydrogen, methane, propane, and gasoline [18]

	Hydrogen	Methane	Propane	Gasoline
Flammability limits in air (vol%) *and lowest ignition energy* (mJ)				
Lower limit	4.0	5.0	2.1	1.0
Backward limit	9.0	5.6	–	–
Upper limit	74.2	15.0	9.4	7.8
Ignition energy	0.019	0.29	0.3	0.24
Auto-ignition temperature (°C)				
Lowest temperature	520	630	450	228-470
Hot gas injection	640	1040	855	–
Ni–Cd heating wire	750	1220	1050	–

Table 1.3 Comparisons of explosion properties between hydrogen, methane, propane, and gasoline [18]

	Hydrogen	Methane	Propane	Gasoline
Explosion limits (vol%) *and maximum laminar burning velocity in air* (m/s)				
Lower limit	13.0–18.3	6.3	3.1	1.1
Upper limit	59.0	13.5	7.0	3.3
Burning velocity	270	37	47	30
Heat of combustion (kJ/g) *and experimental maximum safety gap* (cm)				
Combustion heat	135.4	52.8	40.3	46.0
Max safety gap	0.008	0.12	–	0.074

flammability range and lowest ignition energy. The hydrogen auto-ignition temperature is also relatively low.

In confined spaces, e.g., garages, closed chambers, or tubes/pipes, explosive hydrogen–air mixture can be easily accumulated if there is a leak. Table 1.3 presents comparisons of explosion properties between hydrogen, natural gas, propane, and gasoline [18, 19]. The laminar burning velocity of hydrogen in air is rather high, roughly 8 times and 6 times higher than methane (or gasoline) and propane, respectively. The heat release per unit mass of hydrogen is also the largest. Under the same conditions, hydrogen may be much easier to cause explosion and even DDT. Furthermore, hydrogen flame in air is very difficult to see by naked eyes, so that it is dangerous for one who is close to a hydrogen flame. In addition, flame of hydrogen–air mixture is subjected to instabilities [26, 82], which considerably increases fame speed.

1.2.3.3 Research of Hydrogen–Air Explosion

There have been numerous experimental and numerical studies of hydrogen–air explosions [12, 15, 17, 32, 41, 44, 148–157]. Here, a selected group of the many papers is reviewed, focusing on hydrogen explosions in air relevant to hydrogen safety as an energy carrier.

Early in 1983, Pförtner and Schneider [158] from Fraunhofer Institute (Germany) carried out a series of investigations of hydrogen deflagrations in air with various spatial scales. In their experiments, a premixed stoichiometric hydrogen–air mixture was feed into semi-spheres constructed using polyethylene film with radii from 1.53 to 10.0 m. The mixture was ignited by a spark at the center of the semi-spheres. The experiments showed that the maximum flame propagation speeds during the gas cloud explosions with radii 1.53 and 10.0 m are 43 and 84 m/s, respectively. In order to provide a reliable CFD tool for simulating hydrogen–air explosion, Molkov et al. [152] developed and validated against experiments a large eddy simulation (LES) combustion model based on a RNG (renormalization group theory) and subgrid scale (SGS) approach. Groethe et al. [159] performed experiments of hydrogen–air deflagrations in a semi-sphere with a volume of 300 m^3. The equivalence ratio was 15–30 %. In some of the experiments, eight cylindrical obstacles with diameter of 0.46 m and length of 3 m were set inside the semi-sphere beforehand to clarify if the obstacle could enhance deflagration. The blockage ratio by volume was 11 %. It was found that these obstacles did not increase the propagation speed of the deflagration waves. Royle et al. [160] conducted an experimental work of effect of repeated metal pipes on the vapor cloud explosions of hydrogen–methane–air mixtures in a congested region. The blockage ratio by these pipes was 20 %. The experiments showed that the flame propagation speed and overpressure were significantly increased as the pipe number increased. Nevertheless, no DDT was observed. In juxtaposition to this, Grothe [161] found that the hydrogen–air flame can accelerate drastically and initiate a detonation in regions congested with small pipes. Shell global solutions and UK Health and Safety Laboratory conducted a cooperative experiment of hydrogen–air explosion in an open space [162]. The purpose of the experiment was to develop proper specifications and standards for safety design of hydrogen refueling station. Makarov et al. [148] used this experiment to evaluate the capacity and reliability of several different CFD models for simulating hydrogen–air explosions. This work may be a reference of CFD tools for hydrogen explosion safety.

In addition, Groethe et al. [159] also carried out a series of large-scale experiments in a 78.5-m-long channel with horseshoe-shaped cross section. The size of the channel was about one-fifth of a real tunnel. The height of the channel was 1.84 m and the cross-sectional area was 3.74 m^2. The hydrogen contractions of the premixed hydrogen–air mixtures used were 20 and 30 % by volume. In some of these experiments, a sequence of $940 \times 362 \times 343$ mm^3 obstacles as vehicle models was arranged inside the channel. This experimental work provided fundamental data for studying hydrogen explosion in a tunnel. The data was later used for the evaluation and validation of various CFD models and tools of hydrogen combustion and explosion, such as the work by Molkov et al. [152]. Kumar [163] performed an experimental investigation of flame propagation and pressure build-up in hydrogen–air explosions in a sphere explosion vessel with a diameter of 2.3 m. The initial temperature was 373 K and the initial pressure was slightly lower than pressure of atmosphere. In the experiments, the effect of turbulence induced by a fan on the flame with relatively high hydrogen concentration (hydrogen 27 % and water vapor 10 % by vol.) was examined. The measured turbulent fluctuation velocity was 2 m/s when the fan was turned on the overpressure reached 4 bars within 25 ms, whereas 65 ms was taken to reach the same overpressure when the fan was off. It implies that the external factors, such as fans and other similar devices, can cause more severe explosion by inducing more turbulence. Besides, they also examined the effects of metal grid, ignition location, and hydrogen concentration on the explosions. The data of this work can also be useful for both safety assessment of hydrogen explosions and validation of CFD models and methods. Sato et al. [164] conducted experiments of hydrogen–air explosions in open space and shock tube. They concluded based on these experiments: (1) the overpressure generated in the hydrogen–air explosion with hydrogen concentration 30 % by volume is much higher than that in methane–air explosion with methane concentration 9.5 % by volume (both approximately stoichiometric mixtures), (2) flame propagation speed greatly depends on hydrogen concentration, (3) flame propagation speed and overpressure can be increased by obstacle-induced turbulent flow and increase of the volume of hydrogen–air mixture, (4) hydrogen–air flame in the tube behave in a fast deflagration mode. These experimental results may be used for reexamining current specifications and standards of hydrogen safety.

The interaction of flame with shock waves and DDT in hydrogen explosion is also of importance. Kratzel et al. [39] investigated interactions between a fully developed turbulent flame and a shock wave using high-speed schlieren photography and numerical simulation. Tulip-shaped flame forms in a closed tube under the effect of shock wave, while cellular flame develops in a half-open tube. They thought that the baroclinic effect of shock wave leads to generation of vortices near the flame front. The vortex motion eventually results in changes in flame shape and structure. Chatrathi et al. [165] performed experiments of hydrogen–air flame propagation and DDT in full-size tubes. The effects of various parameters, e.g., mixture composition, tube diameter, and curved section, on explosions were studied.

1.3 Scientific Issues and Research Objectives of the Thesis

From above review of research on premixed flame dynamics in confined regions and hydrogen–air explosions, we know the following:

(1) Premixed flame dynamics of combustible gas is an important, fundamental subject of combustion research. Premixed flame propagation and pressure

build-up are typical processes occurring during gas explosions. There have been a large number of studies of flame acceleration, turbulent flame, DDT, and cellular detonation in tubes. Nevertheless, the flame dynamics has not been sufficiently explored. And it is necessary to further study the early flame dynamics in tubes since it usually represent the initial flame acceleration and therefore play an essential role in gas explosion process in confined regions.

(2) Although significant advances have been made in experimental, numerical, and theoretical research of gas explosions in tubes, the flame propagation phenomena and mechanisms have not been sufficiently understood. Besides, there is a lack of theoretical models for entire flame evolution in tubes, particularly when the flame undergoes complex shape changes or develops turbulent regime. With respect to numerical simulation, it is difficult to accurately capture the flame acceleration and DDT under different conditions. Gas explosion in a tube is a transient chemically reacting process, involving compressible flow, turbulence, and chemical reaction. Turbulent flow remains an unresolved issue in scientific community. It is computationally expensive to couple detailed chemistry mechanism in the calculation of moderate/large-scale explosions since the differences in length and time scales can be extremely large. In these cases, combustion modeling is important.

(3) Previous studies of flame propagation were mainly focused on flame acceleration and DDT phenomena, as well as validation of numerical models and approaches. Flame propagation speed and overpressure are two important parameters in these works. The detailed flame dynamics and effects of various factors, such as boundary layer, pressure wave, and gravity, need to be further examined. These are also very important for both revealing mechanisms controlling flame propagation and validating numerical models.

(4) Flame instabilities play an important role in the flame dynamics in tubes. The linear instabilities are crucial for the initiation of flame shape changes, while the nonlinear instabilities determine the development and consequence of flame deformation. However, to the author's best knowledge, a nonlinear theory of flame instabilities may have not been fully developed so far.

(5) The interactions between flame, pressure waves, and combustion-generated flow have significant influences on flame dynamics in tubes. It is usually difficult to experimentally measure these interactions during explosions which involve high-speed reacting flow. CFD computations can help to gain details of these interactions, but only for rather small-scale problems. Fortunately, thanks to the rapid development in both computer performance and numerical techniques, CFD simulation will, in future, solve problems of scales that seem currently prohibited.

(6) Safety of hydrogen as an energy carrier involves hydrogen leak, ignition, fire, and explosion. Combustion and explosion of hydrogen–air mixtures have been one of the essential research fields of hydrogen safety and combustion application. However, there is still huge gap in the knowledge of hydrogen combustion and explosions, especially under practical or extreme conditions.

Overall, there remain numerous scientific issues to be resolved concerning both premixed flame dynamics in confined regions and hydrogen explosion safety. The purpose of this thesis is to conduct experimental, numerical, and theoretical research of the behavior, characteristics, and underlying mechanisms of hydrogen flames propagating in tubes. In the experimental research, high-speed schlieren photography and pressure transducer are used to record the flame structures, flame front dynamics, and pressure build-up during the hydrogen–air flame propagation under various conditions. The experimental work also aims to present important knowledge of the effects of various parameters on flame dynamics, e.g., equivalence ratio, opening ratio, and gravity. The numerical research will help to elucidate the flame mechanisms and give a deep insight into the interactions of flame with the combustion-generated flow, pressure wave, and boundary layer. Based on the experimental and numerical studies, analytical investigations of premixed hydrogen–air flame evolution in a tube will be provided to suggest a theory of hydrogen flame dynamics in a closed tube. Efforts will also be made in the dissertation to validate numerical approaches for simulating premixed flame propagation in hydrogen–air explosions. The analysis and evaluation of different combustion approaches would also facilitate our understanding of the influences of physical mechanisms and numerical methods on the flame dynamics.

1.4   Research Content and Organization of the Thesis

The research content of the present thesis is as follows:

(1) Experimentally investigate the fundamental phenomena of combustion dynamics of hydrogen–air mixtures in ducts using high-speed schlieren photography and pressure transducer, including changes in flame shape and position as a function of time, pressure dynamics, relationships between flame dynamics, and pressure waves. Reveal the effects of equivalence ratio, opening ratio of tube, gravity, and other factors on the flame dynamics.

(2) Numerically study hydrogen–air flame structure (species profiles and flame thickness) and laminar burning velocity which are important fundamental parameters for the flame dynamics.

(3) Conduct numerical simulations of premixed hydrogen–air flames propagating in closed tubes using thickened flame (TF) model and burning velocity model along with LES. Then compare the results of these simulations with the experimental measurements. Study the interactions between flame, flow, and pressure waves based on the numerical simulations and subsequently reveal the mechanisms underlying the flame propagation. Examine the flow and combustion regimes. The effect of wall friction will be also studied.

(4) Suggest and validate theoretical model and CFD approaches of premixed hydrogen–air flame propagation in tubes based on experimental results.

The thesis consists of six chapters. In this chapter, we provide a background and an introduction of the research, and then give the objectives, content, and organization of the thesis. In Chapter 2, we first describe experimental setup, devices, and procedures, and then systematically study premixed hydrogen–air flames propagating in ducts using the experimental apparatus. In Chapter 3, we first present the CFD approaches and schemes used in the research, and then describe the numerical investigation of premixed hydrogen–air flame propagation in closed ducts. In Chapter 4, we conduct theoretical analysis of premixed hydrogen–air flame propagation in a closed tube on the basis of experimental and numerical results. In Chapter 5, we discuss the interactions between flame front, combustion-generated flow, and pressure waves to gain an insight into the mechanisms leading to curious phenomena of flame instabilities. Summary, conclusions, and recommendations of future research are given in Chapter 6.

References

1. Bjerketvedt D, Bakke JR, van Wingerden K (1997) Gas explosion handbook. J Hazard Mater 52:1–150
2. Dorofeev SB (2011) Flame acceleration and explosion safety applications. Proc Combust Inst 33:2161–2175
3. Liberman M (2003) Flame, detonation, explosion—when, where and how they occur. In: Third international disposal conference, Karlskoga, Sweden, pp 5–23
4. Lea CJ, Ledin HS (2002) A review of the state-of-the-art in gas explosion modeling. In: Health & safety Laboratory, fire and explosion group, Hampur Hill, Buxton, SK179JN
5. Kristoffersen K (2004) Gas explosions in process pipes. Telemark University College, Porsgrunn, Norway
6. Pande JO, Tonheim J (2001) Ammonia plant NII: explosion hydrogen in a pipeline for CO2. Process Saf Prog 20:37–40
7. Kinsmann P, Lewis J (2002) Report on a second study of pipeline accidents using the health and safety executive's risk assessment programs MISHAP and PIPERS. Health and Safety Executive [HSE] Research Report, 063, 2002
8. Bi M (2001) A research on the pressure fields of unconfined flammable gas cloud explosions. Dalian University of Technology, Dalian (in Chinese)
9. Wang Z (2005) Study on the dynamics of gas explosion process in confined space. Nanjing University of Technology, Nanjing (in Chinese)
10. Hirano T (2004) Accident explosions of semiconductor manufacturing gases in Japan. J Loss Prev Process Ind 17:29–34
11. Bematik A, Libisova M (2004) Loss prevention in heavy industrial: risk assessment of large gasholder. J Loss Prev Process Ind 17:271–278
12. Ng HD, Lee JHS (2008) Comments on explosion problems for hydrogen safety. J Loss Prev Process Ind 21:136–146
13. Fairweather M, Hargrave GK, Ibrahim SS, Walker DG (1999) Studies of premixed flame propagation in explosion tubes. Combust Flame 116:504–518
14. Gamezo VN, Ogawa T, Oran ES (2008) Flame acceleration and DDT in channels with obstacles: effect of obstacle spacing. Combust Flame 155:302–315
15. Dorofeev SB (2009) Hydrogen flames in tubes: critical run-up distances. Int J Hydrogen Energy 34:5832–5837
16. Law CK (2007) Combustion at a crossroads: status and prospects. Proc Combust Inst 31:1–29

17. Molkov V. Hydrogen safety research: state-of-the-art. Proceedings of the 5th International Seminar on Fire and Explosion Hazards. Edinburgh, UK. 2007: 28-43

18. Feng W, Wang S, Ni W, Chen C (2003) The safety of hydrogen energy and fuel cell vehicles. Acta Energiae Solaris Sinica 24:678–682 (in Chinese)

19. Green MA (2006) Hydrogen safety issues compared to safety issues with methane and propane. In: Proceedings of the Advances in cryogenic engineering, vols 51A and B, F

20. Jacob PO (2005) DEO guidelines on hydrogen safety. In: SAE World Congress

21. HySafe. Safety of hydrogen as an energy carrier, electronic information. http://www.hysafe.org/

22. HySafe (2005) HySafe, HySafe 1st Periodic report, executive summary. http://www.hysafe.org/

23. Hirschfelder JO, Curtis CF (1949) Theory of propagation of flames. In: Wilkins WA (ed) Third symposium on combustion, flame and explosion phenomena, Baltimore, p 121

24. Hirschfelder JO, Curtis CF, Bird RB (1954) Molecular theory of gases and liquids. New York

25. Markstein GH (1964) Nonsteady flame propagation. Pergamon Press Limited, New York

26. Matalon M (2009) Flame dynamics. Proc Combust Inst 32:57–82

27. Xiao H, Wang Q, He X, Sun J, Shen X (2011) Experimental study on the behaviors and shape changes of premixed hydrogen-air flames propagating in horizontal duct. Int J Hydrogen Energy 36:6325–6336

28. Xiao H, Makarov D, Sun J, Molkov V (2012) Experimental and numerical investigation of premixed flame propagation with distorted tulip shape in a closed duct. Combust Flame 159:1523–1538

29. Poinsot T, Veynante D (2005) Theoretical and numerical combustion, 2nd edn. Edwards RT Inc, Philadelphia

30. Ciccarelli G, Dorofeev S (2008) Flame acceleration and transition to detonation in ducts. Prog Energy Combust Sci 34:499–550

31. Verhelst S, Wallner T (2009) Hydrogen-fueled internal combustion engines. Prog Energy Combust Sci 35:490–527

32. Dorofeev SB, Kuznetsov MS, Alekseev VI, Efimenko AA, Breitung W (2001) Evaluation of limits for effective flame acceleration in hydrogen mixtures. J Loss Prev Process Ind 14:583–589

33. Oran ES, Boris JP (2000) Numerical simulation of reactive flow. Cambridge University Press, Cambridge

34. Abu-Orf GM, Cant RS (2000) A turbulent reaction rate model for premixed turbulent combustion in spark-ignition engines. Combust Flame 122:233–252

35. Akkerman V, Bychkov V, Petchenko A, Eriksson LE (2006) Accelerating flames in cylindrical tubes with nonslip at the walls. Combust Flame 145:206–219

36. Bychkov V, Akkerman V, Fru G, Petchenko A, Eriksson LE (2007) Flame acceleration in the early stages of burning in tubes. Combust Flame 150:263–276

37. Bychkov VV, Liberman MA (2000) Dynamics and stability of premixed flames. Phys Rep 325:116–237

38. Kagan L, Sivashinsky G (2003) The transition from deflagration to detonation in thin channels. Combust Flame 134:389–397

39. Kratzel T, Pantow E, Fischer M (1998) On the transition from a highly turbulent curved flame into a tulip flame. Int J Hydrogen Energy 23:45–51

40. Law CK (2006) Combustion physics. Cambridge University Press, New York

41. Makarov DV, Molkov VV (2004) Modeling and large eddy simulation of deflagration dynamics in a closed vessel. Combust Explos Shock Waves 40:136–144

42. Matalon M, Mcgreevy JL (1994) The initial development of a tulip flame. In: Proceedings of the Combustion Institute, pp 1407–1413

43. Matalon M, Metzener P (1997) The propagation of premixed flames in closed tubes. J Fluid Mech 336:331–350

44. Molkov V, Makarov D, Grigorash A (2004) Cellular structure of explosion flames: Modeling and large-eddy simulation. Combust Sci Technol 176:851–865
45. Starke R, Roth P (1986) An experimental investigation of flame behavior during cylindrical vessel explosions. Combust Flame 66:249–259
46. Starke R, Roth P (1989) An experimental investigation of flame behavior during explosions in cylindrical enclosures with obstacles. Combust Flame 75:111–121
47. Xiao HH, Wang QS, He XC, Sun JH, Yao LY (2010) Experimental and numerical study on premixed hydrogen/air flame propagation in a horizontal rectangular closed duct. Int J Hydrogen Energy 35:1367–1376
48. Dunn-Rankin D, Barr PK, Sawyer RF (1986) Numerical and experimental study of "tulip" flame formation in a closed vessel. Proc Combust Inst 21:1291–1301
49. Dunn-Rankin D, Sawyer RE (1985) Interaction of a laminar flame with its self-generated flow during constant volume combustion. In: Proceedings of the 10th ICDERS, Berkley, California
50. Dunn-Rankin D, Sawyer RF (1998) Tulip flames: changes in shape of premixed flames propagating in closed tubes. Exp Fluids 24:130–140
51. Mogi T, Kim D, Shiina H, Horiguchi S (2008) Self-ignition and explosion during discharge of high-pressure hydrogen. J Loss Prev Process Ind 21:199–204
52. Dryer FL, Chaos M, Zhao Z, Stein JN, Alpert JY, Homer CJ (2007) Spontaneous ignition of pressurized releases of hydrogen and natural gas into air. Combust Sci Technol 179:663–694
53. Glassman I (1987) Combustion, 2nd edn. Academic press inc., San Diego
54. Williams FA (1985) Combustion theory. Perseus Books, Massachusetts
55. Bradley D, Gaskell PH, Gu XJ (1996) Burning velocities, markstein lengths, and flame quenching for spherical methane-air flames: a com-putational study. Combust Flame 104:176–198
56. Lamoureux N, Djebaili-Chaumeix N, Paillard CE (2003) Laminar flame velocity determination for H2-air-He-CO2 mixtures using the spherical bomb method. Exp Thermal Fluid Sci 27:385–393
57. Huang ZH, Zhang Y, Zeng K, Liu B, Wang Q, Jiang D (2006) Measurements of laminar burning velocities for natural gas-hydrogen-air mixtures. Combust Flame 146:302–311
58. Hu E, He J, Huang Z, Jin C, Wang J, Miao H, Wang T, Jiang D (2008) Measurement of laminar burning velocities and flame instabilities of diluted hydrogen-air mixtures. Chin Sci Bull 53:2514–2525 (in Chinese)
59. Fristrom RM, Westenberg AA (1965) Flame structure. Mc Graw Hill, New York
60. Kee RJ, Rupley FM, Miller JA (1989) Technical report SAND89-8009. Sandia National Laboratories
61. Teerling OJ (2004) Acoustics and flame acceleration in regions of partial confinement. University of Leeds, Leeds, UK
62. Aung KT, Hassan MI, Faeth GM (1997) Flame stretch interactions of laminar premixed hydrogen/air flames at normal temperature and pressure. Combust Flame 109:1–24
63. Landau LD, Lifshitz EM (1987) Fluid mechanics. Pergamon Press, Oxford
64. Kwon OC, Faeth GM (2001) Flame/stretch interactions of premixed hydrogen-fueled flames: measurements and predictions. Combust Flame 124:590–610
65. Jiao Q, Miao H, Huang Q, Huang Z, Jiang D, Zeng K (2009) Influence of initial pressure on flame propagation characteristics of natural gas-hydrogen-air mixture. J Combust Sci Technol 15:374–380 (in Chinese)
66. Aung KT, Hassan MI, Faeth GM (1998) Effects of pressure and nitrogen dilution on flame/stretch interactions of laminar premixed H2/O2/N2 flames. Combust Flame 112:1–15
67. Liu FS, Guo HS, Smallwood GJ, Gulder OL (2002) Numerical study of the superadiabatic flame temperature phenomenon in hydrocarbon premixed flames. Proc Combust Inst 29:1543–1550
68. Dong C, Zhou Q, Zhao Q, Zhang Y, Xu T, Hui S (2009) Experimental study on the laminar flame speed of hydrogen/carbon monoxide/air mixtures. Fuel 88:1858–1863

69. Dahoe AE (2005) Laminar burning velocities of hydrogen-air mixtures from closed vessel gas explosions. J Loss Prev Process Ind 18:152–166
70. Law CK (1993) A compilation of experimental data on laminar burning velocities, reduced kinetic mechanisms for applications in combustion systems. Springer, pp 15–26
71. Takahashi F, Mizomoto M, Ikai S (1983) Nuclear energy/synthetic fuels. McGraw-Hill, New York
72. Wu CK, Law C (1984) On the determination of laminar flame speeds from stretched flames. In: Proceedings of the Combustion Institute, 1941–1949
73. Iijima T, Takeno T (1986) Effects of temperature and pressure on burning velocity. Combust Flame 65:35–43
74. Dowdy DR, Smith DB, Taylor SC, Williams A (1990) The use of expanding spherical flames to determine burning velocities and stretch effects in hydrogen-air mixtures. In: Proceedings of the Combustion Institute, pp 325–332
75. Egolfopoulos FN, Law CK (1990) An experimental and computational study of the burning rates of ultra-lean to moderately rich H2/O2/N2 laminar flames with pressure variations. In: Proceedings of the Combustion Institute, pp 333–346
76. Koroll GW, Kumar RK, Bowles EM (1993) Burning velocities of hydrogen-air mixtures. Combust Flame 94:330–340
77. Vagelopoulos CM, Egolfopoulos FN. Further considerations on the determination of laminar flame speeds from streched flames. In: Proceedings of the Combustion Institute, pp 1341–1347
78. Tse SD, Zhu DL, Law CK (2000) Morphology and burning rates of expanding spherical flames in H2/O2/inert mixtures up to 60 atmospheres. Proc Combust Inst 28:1793–1800
79. Lamoureux N, Djebaili-Chaumeix N, Paillard CE (2002) Laminar flame velocity determination for H2-air-steam mixtures using the spherical bomb method. J Phys France IV 12:445–452
80. Law CK (2006) Propagation, structure, and limit phenomena of laminar flames at elevated pressures. Combust Sci Technol 178:335–360
81. Nagy J, Conn J, Verakis H (1969) Explosion development in a spherical vessel. Technical Reports RI 7279, US Department of Interior, Bureau of Mines
82. Day M, Bell J, Bremer PT, Pascucci V, Beckner V, Lijewski M (2009) Turbulence effects on cellular burning structures in lean premixed hydrogen flames. Combust Flame 156:1035–1045
83. Chen Z, Ju Y, Minaev S (2002) Flammability and stability analysis of cylindrical flames. J Eng Thermophys 23:513–516 (in Chinese)
84. Darrieus G (1938) Proceedings of the propagation d'un front de flamme, presented at La Technique Moderne, Paris, F, 1938 (unpublished work)
85. Landau LD (1944) On the theory of slow combustion. Acta PhyscoChim USSR 19:77–85
86. Lipatnikov AN, J. C. Molecular transport effect on turbulent flame propagation and structure. Progr Energy Combust Sci 31:1–7
87. Sivashinsky GI (1979) On self-turbulization of a laminar flame. Acta Astronaut 5:569–591
88. Dorofeev S (2008) Flame acceleration and transition to detonation: a framework for estimating potential explosion hazards in hydrogen mixtures. In: Lecture presented at the 3rd European Summer School on hydrogen safety, Belfast, UK
89. Teodorczyk A, Lee JHS (1995) Detonation attenuation by foams and wire meshes lining the walls. Shock Waves 4:225–236
90. Liu F, McIntosh AC, Brindley J. A numerical investigation of Rayleigh-Taylor effects in pressure wave-premixed flame interaction. Combust Sci Technol 91:373–386
91. Bradley D, Harper C (1994) The development of instabilities in laminar explosion flames. Combust Flame 99:562–572
92. Zeldovich YB, Barenblatt GI, Librovich VB, Makhviladze GM (1985) The mathematical theory of combustion and explosions. Consultants Bureau, New York
93. Wang F (2004) Computational fluid dynamics analysis: the principle and application of CFD software. Tsinghua University Press, Beijing (in Chinese)
94. Pope SB (2000) Turbulent flows. Cambridge University Press, Cambridge
95. Borghi R (1985) Recent advances in aerospace science. Plenum Press, New York

96. Warnatz J, Maas U, Dibble RW (2006) Combustion: physical and chemical fundamentals, Modeling and simulation, experiments, pollutant formation, 4th edn. Springer-Verlag, Berlin Heidelberg, Berlin, Germany

97. Peters N (2000) Turbulent combustion. Cambridge University Press, Cambridge

98. Chapman DL (1899) On the rate of explosion in gases. Philos Mag 47:90–104

99. Jouguet E (1905) On the propagation of chemical reactions in gases. J Math Pures Appl 6:347

100. Zeldovich YB (1940) On the theory of the propagation of detonation in gaseous systems. Zhur Eksp Teor Fiz 10:542–568

101. von Neumann J (1942) Theory of detonation waves. OSRD Report, No. 549

102. Döring W (1943) On detonation processes in gases. Ann Phys 43:421–436

103. Denisov YN, Troshin YK (1960) On the mechanism of detonative combustion. In: Proceedings of the Combustion Institute, pp 600–610

104. Denisov YN, Troshin YK (1960) Structure of gaseous detonations in channels. Zeitschrift fur technische Physik 30:450

105. Han Q, Wang J, Wang B (2003) Investigation of deflagration to detonation transition distance in a tube with mixture. J Propul Technol 24:63–66 (in Chinese)

106. Xia C, Zhou K, Shen Z, Dong Y, Nian W, Wang H (2002) Exprimental study on propagation characteristics of unsteady gaseous detonation through 90° round bend. J Exp Mech 17:438–443 (in Chinese)

107. Yu L, Sun W, Wu C (2002) Flame propagation of H2-air in a semi-open obstructed tube. J Combust Sci Technol 8:27–30 (in Chinese)

108. Mallard M, Le Chatelier H (1883) Recherches experimentales et theoriques sur la combustion des m´elanges gazeux explosifs. Annales de Mines 8:274–568

109. Ellis OCde C (1928) Flame movement in gaseous explosive mixtures. Fuel Sci 7:502–508

110. Salamandra GD, Bazhenova TV, Naboko IM (1959) Formation of detonation wave during combustion of gas in combustion tube. Proc Combust Inst 7:851–855

111. Gonzalez M, Borghi R, Saouab A (1992) Interaction of a Flame Front with Its Self-Generated Flow in an Enclosure—the Tulip Flame Phenomenon. Combust Flame 88:201–220

112. Marra FS, Continillo G (1996) Numerical study of premixed laminar flame propagation in a closed tube with a full Navier-Stokes approach. Proc Combust Inst 26:907–913

113. Metzener P, Matalon M (2001) Premixed flames in closed cylindrical tubes. Combust Theor Model 5:463–483

114. Clanet C, Searby G (1996) On the "tulip flame" phenomenon. Combust Flame 105:225–238

115. Schmidt EHW, Steinecke H, Neubert U (1952) Flame and schlieren photographs of combustion waves in tubes. Proc Combust Inst 4:658–666

116. Gonzalez M (1996) Acoustic instability of a premixed flame propagating in a tube. Combust Flame 107:245–259

117. Rotman DA, Oppenheim AK (1986) Aerothermodynamic properties of stretched flames in enclosures. In: Proceedings of the Combustion Institute, pp 1303–1310

118. Markstein GH (1956) A shock-tube study of flame front-pressure wave interaction. Proc Combust Inst 6:387–398

119. Barrere M, Williams FA (1976) Comparison of combustion instabilities found in various types of combustion chambers. Proceedings of the Combustion Institute, pp 169–181

120. Guenoche H (1964) Flame propagation in tubes and in closed vessels. In: Markstein GH (ed) Nonsteady flame propagation. Pergamon Press, New York, p 107

121. Chen X (2007) Study on fine flame structure behavior and flame accelerating mechanism of premixed propane-air. University of Science and Technology of China, Hefei (in Chinese)

122. Ye J, Jia Z, Dong G, Fan B (2005) Experiment and theory investigated of shock-flame interactions. J Eng Thermophys 26:511–514 (in Chinese)

123. Fan B, Jiang Q, Dong G, Ye J (2003) The time evolution of shock-flame interaction. Explos Shock Waves 23:488–492 (in Chinese)

124. Evens MW, Scheer MD, Schoen LJ, Miller JA (1948) A study of high velocity flames developed by grids in tubes. In: Proceedings of the Combustion Institute, pp 168–176

125. Catlin CA, Fairweather M, Ibraiim SS (1995) Predictions of turbulent, premixed flame propagation in explosion tubes. Combust Flame 102:115–128
126. Gubba SR, Ibrahim SS, Malalasekera W, Masri AR (2011) Measurements and LES calculations of turbulent premixed flame propagation past repeated obstacles. Combust Flame 158:2465–2481
127. Kessler DA, Gamezo VN, Oran ES (2010) Simulations of flame acceleration and deflagration-to-detonation transitions in methane-air systems. Combust Flame 157:2063–2077
128. Gamezo VN, Ogawa T, Oran ES (2007) Numerical simulations of flame propagation and DDT in obstructed channels filled with hydrogen-air mixture. Combust Flame 31:2463–2471
129. Lee JH, Knystautas R, Chan CK (1984) Turbulent flame propagation in obstacle filled tubes. In: Proceedings of the Combustion Institute, pp 1663–1672
130. Kuznetsov M, Alekseev V, Yankin Y, Dorofeev S (2002) Slow and fast deflagrations in hydrocarbon-air mixtures. Combust Sci Technol 174:157–172
131. Masri AR, Ibrahim SS, Nehzat N, Green AR (2000) Experimental study of premixed flame propagation over various solid obstructions. Exp Thermal Fluid Sci 21:109–117
132. Ibrahim SS, Masri AR (2001) The effects of obstructions on overpressure resulting from premixed flame deflagration. J Loss Prev Process Ind 14:213–221
133. Patel SNDH, Jarvis S, Ibrahim SS, Hargrave GK (2002) An experimental and numerical investigation of premixed flame deflagration in a semiconfined explosion chamber. Proc Combust Inst 29:1849–1854
134. Johansen T, Ciccarelli G (2009) Visualization of the unburned gas flow field ahead of an accelerating flame in an obstructed square channel. Combust Flame 156:405–416
135. Zhou K, Li Z (2000) Flame front acceleration of prone-air deflagration in straight tubes. Explos Shock Waves 20:137–142 (in Chinese)
136. Lin B, Zhang R, Lu H (1999) Research on accelerating mechanism and flame transmission in gas explosion. J China Coal Soc 24:56–59 (in Chinese)
137. He X, Yang Y, Wang E, Liu Z (2004) Effects of obstacle on premixed flame microstructure and flame propagation in methane/air explosion. J China Coal Soc 29:186–189 (in Chinese)
138. Kuznetsov M, Alekseev V, Matsukov I, Dorofeev S (2005) DDT in a smooth tube filled with a hydrogen-oxygen mixture. Shock Waves 14:205–215
139. Zeldovich YB, Librovich VB, Makhviladze GM, Sivashinsky GI (1970) On the development of detonation in a non-uniformly preheated gas. Astronaut Acta 15:313–321
140. Dorofeev SB (2002) Flame acceleration and DDT in gas explosions. J Phys 12:3–10
141. Zhou B, Sobiesiak A, Quan P (2006) Flame behavior and flame-induced flow in a closed rectangular duct with a 90 degrees bend. Int J Therm Sci 45:457–474
142. Sato Y, Sakai Y, Chiga M (1996) Flame propagation along 90° bend in an open duct. Proc Combust Inst 26:931–937
143. Wang C, Xu S, Guo C (2003) Experimental investigation on gaseous detonation propagation through a semi-circle bend tube. Explos Shock Waves 23:448–453 (in Chinese)
144. Wang C, Guo C, Xu S, Zhang H (2004) Experimental investigation on gaseous detonation propagation through a T-shape bifurcated tube. Acta Mech Sin 36:16–23 (in Chinese)
145. Wang C, Xu S, Fei L, Guo C (2006) Schlieren visualization and numerical simulation on gaseous detonation propagation through a bend tube. Acta Mech Sin 38:9–15 (in Chinese)
146. He X (2009) Experimental and Numerical study on Characteristics of premixed propane-air flame in a rectangular duct with a 90° bend. University of Science and Technology of China, Hefei (in Chinese)
147. Company FM (1997) Direct-hydrogen-fueled proton-exchange-membrane fuel cell system for transportation applications: hydrogen vehicle safety report (DE-AC02294CE50389). US Department Of Energy, 1997
148. Makarov D, Verbecke F, Molkov V, Roe O, Skotenne M, Kotchourko A, Lelyakin A, Yanez J, Hansen O, Middha P, Ledin S, Baraldi D, Heitsch M, Efimenko A, Gavrikov A (2009) An inter-comparison exercise on CFD model capabilities to predict a hydrogen explosion in a simulated vehicle refuelling environment. Int J Hydrogen Energy 34:2800–2814

149. Makarov DV, Molkov VV (2004) Large eddy simulation of gaseous explosion dynamics in an unvented vessel. Combust Explos Shock Waves 40:136–144
150. Molkov V (2009) A multiphenomena turbulent burning velocity model for large eddy simulation of premixed combustion. In: Roy GD (2009) Nonequilibrium phenomena: plasma, combustion, atmosphere. Moscow; Torus Press Ltd. 2009: 315-323
151. Molkov V, Makarov D, Puttock J (2006) The nature and large eddy simulation of coherent deflagrations in a vented enclosure-atmosphere system. J Loss Prev Process Ind 19:121–129
152. Molkov V, Makarov D, Schneider H (2006) LES modelling of an unconfined large-scale hydrogen-air deflagration. J Phys D Appl Phys 39:4366–4376
153. Molkov V, Verbecke F, Makarov D (2008) LES of hydrogen-air deflagrations in a 78.5-m tunnel. Combust Sci Technol 180:796–808
154. Molkov VV, Makarov DV, Schneider H (2007) Hydrogen-air deflagrations in open atmosphere: Large eddy simulation analysis of experimental data. Int J Hydrogen Energy 32:2198–2205
155. Baraldi D, Kotchourko A, Lelyakin A, Yanez J, Gavrikov A, Efimenko A, Verbecke F, Makarov D, Molkov V, Teodorczyk A (2010) An inter-comparison exercise on CFD model capabilities to simulate hydrogen deflagrations with pressure relief vents. Int J Hydrogen Energy 35:12381–12390
156. Baraldi D, Kotchourko A, Lelyakin A, Yanez J, Middha P, Hansen OR, Gavrikov A, Efimenko A, Verbecke F, Makarov D, Molkov V (2009) An inter-comparison exercise on CFD model capabilities to simulate hydrogen deflagrations in a tunnel. Int J Hydrogen Energy 34:7862–7872
157. Wen JX, Madhav RVC, Tam VHY (2010) Numerical study of hydrogen explosions in a refuelling environment and in a model storage room. Int J Hydrogen Energy 35:385–394
158. Pförtner H, Schneider H (1983) Fraunhofer-institut fur treib-und explosivstoffe. ICT-Projektforschung 19/83 Forschungsprogramm "Prozeßgasfreisetzung—Explosion in der Gasfabrik und Auswirkungen von Druckwellen auf das Containment" Ballonversuche zur Untersuchung der Deflagration von Wasserstoff/Luft-Gemischen (Abschlußbericht)
159. Groethe M, Merilo E, Sato Y (2005) Large-scale hydrogen deflagrations and detonations. In: International conference on hydrogen safety, Pisa, Italy
160. Royle M, Shirvill LC, Roberts TA (2007) Vapour cloud explosions from the ignition of methane/hydrogen/air mixtures in a congested region. In: International conference on hydrogen safety, San Sebastian
161. Groethe M (2012) Hydrogen deflagration safety studies in a semi-open space. In: 14th World hydrogen energy conference, Montreal, Quebec, Canada
162. Shirvill LC, Roberts TA, Royle M (2005) Safety studies on high-pressure hydrogen vehicle refueling stations: part 1 Releases into a simulated high-pressure dispensing area. Confid Draft Hysafe
163. Kumar RK (1983) Combustion of hydrogen at high concentrations including the effects of obstacles. Combust Sci Technol 35:175–186
164. Sato Y, Iwabuchi H, Groethe M, Merilo E, Chiba S (2006) Experiments on hydrogen deflagration. J Power Sources 159:144–148
165. Chatrathi K, Going JE, Grandestaff B (2001) Flame propagation in industrial scale piping. Process Saf Progr 20:286–294

Chapter 2
Experiments of Premixed Hydrogen–Air Flame Propagation in Ducts

2.1 Introduction

Experimental study is essential for revealing new phenomena of flame propagation and developing new theory and numerical models [1]. For flame propagation in tubes, most of the previous studies primarily focused on flame acceleration and DDT [2–5] while fewer studies on detailed flame structure and shape changes [6]. The usual combustible gases used were hydrocarbons such as methane, propane, and ethylene [7–11]. Although hydrogen–air flames may differ from those of hydrocarbons [12], the study of premixed hydrogen–air flame propagation in tubes is not enough. With emerging hydrogen economy, large-scale experiments of hydrogen–air explosions in open spaces and tunnels were carried out [12–18], aiming to provide data for safety evaluation and validation of numerical models. In these works, primary attention was paid to explosion dynamics parameters, namely overpressure, pressure rise rate, flame speed, and the influence of obstacles arranged beforehand. The information on detailed premixed hydrogen–air flame shape changes, characteristics, and dynamics included in this work could also be used to assess the accuracy and robustness of the theoretical and numerical models/assumptions.

In this experimental study, to further reveal premixed hydrogen–air flame behaviors, characteristics, and mechanisms, and to elucidate the effects of equivalence ratio, opening area, and gravity on flame dynamics, an experimental system of premixed flame propagation is constructed and a series of experiments are conducted.

2.2 Experimental Setup and Methods

The experimental setup is shown in Fig. 2.1. It consists of a constant volume combustion vessel, a high-speed schlieren cinematography system, a pressure recording system, a gas mixing system, a high-voltage ignition system, and a synchronization controller.

© Springer-Verlag Berlin Heidelberg 2016
H. Xiao, *Experimental and Numerical Study of Dynamics
of Premixed Hydrogen-Air Flames Propagating in Ducts*,
Springer Theses, DOI 10.1007/978-3-662-48379-4_2

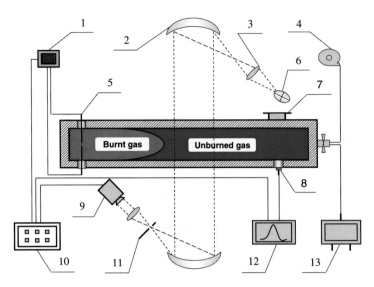

Fig. 2.1 Sketch of experimental apparatus: (*1*) spark igniter, (*2*) schlieren mirror, (*3*) focusing lens, (*4*) vacuum pump, (*5*) ignition electrode, (*6*) point light source, (*7*) discharge vent, (*8*) pressure transducer, (*9*) high-speed video camera, (*10*) synchronization controller, (*11*) knife edge, (*12*) data recorder, (*13*) gas mixing device

2.2.1 Combustion Tube

The combustion vessel is a rectangular duct 82 mm square by 530 mm long. Parallel side walls are important for schlieren photography and there is little qualitative difference between tulip formation in square cross-section and circular cross-section tubes [11]. The duct is placed horizontally in the center of the schlieren optical path. The two side panels of the duct are made of quartz glass with a thickness of 1.6 cm to provide optical access, while the upper and lower walls are made of TP304 stainless steel with a thickness of 1.5 cm. An ignition electrode is placed at the duct axis at a distance of 5.5 cm from the left-end wall. A mounting hole is located at the bottom of the duct for installation of pressure transducer, at a distance of 7.5 cm from the right-end wall. Two valves are set in the right-end wall for gas filling and tube vacuuming. A discharge vent, which is initially closed, is setup close to the right end of the duct for safety. The vent is a circle with a diameter of 4.0 cm and its center is located at a distance of 7.5 cm from the right-end wall.

2.2.2 Gas Mixture Preparation and Filling System

The gas mixture preparation and filling system is schematically shown in Fig. 2.2. The mixture is a combination of pure hydrogen and dry air. The hydrogen

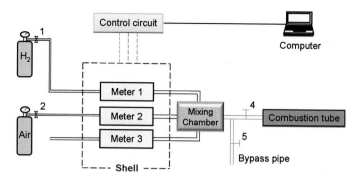

Fig. 2.2 Schematic of hydrogen–air mixture preparation and filling system

concentration is controlled using high-precision mass flow meters 1 and 2 in the experiments. Note that mass flow meter 3 is designed for future experiments on effect of addition of inhibitor gases on flame dynamics.

Before mixture preparation, the system should run at least 20 min to warm up itself. When preparing the mixture, first set the hydrogen concentration and mass flow rate in the computer, then open valves 1, 2, and 4 and close valve 3. The pressure gauges of hydrogen and air gas cylinders 1 and 2 should be kept around 2.5 atm during the gas preparation process. Hydrogen and air are mixed in the premixing chamber. The system needs approximately 1 min to stabilize the hydrogen concentration, and the equivalence ratio of the prepared mixture within the first 1 min may differ from the target ratio. Thus the mixture at the first 1 min should be discharged into open air through valve 4 of the bypass pipe. After that, close valve 4 and open valve 3. The premixed hydrogen–air mixture enters into the combustion tube. When the pressure inside the tube reaches 1 atm, stop filling. When a set of tests are finished, valves 1 and 2 need to be closed to prevent gas leak. If the tube is filled with mixture by purging, the purging time is 3–5 min for fuel-lean mixture and 5–10 min for fuel-rich mixture.

2.2.3 High-Voltage Ignition System

In laboratory and industry, there are various ignition techniques, e.g., chemical ignition, high-voltage electrode, spark, compression, and laser. Chemical ignition source such as priming can be very strong, and it is usually used for engineering blasting or detonation initiation. Most of the explosion accidents in industries are due to weak ignition source. In order to mimic general explosion process in industries, a weak ignition technique, i.e., capacitor discharging, is adopted. A home-designed high-voltage pulse generator Model 2002-1 is applied to charge the capacitor by normal alternating current. When discharging, the voltage is increased by a ratio of 50:1. A transient high voltage is imposed to the electrodes which subsequently create spark for ignition.

The spark energy released by the high-voltage electrodes depends on the stored energy in the capacitor [19]:

$$E = \frac{1}{2}CU^2,\tag{2.1}$$

where E is the stored energy in the capacitor (J), $C = 200.0$ μF is the capacitance, and $U = 300.0$ V is the charging voltage. The stored energy calculated by Eq. (2.1) is 9.0 J. The ignition system is triggered by the synchronization device.

2.2.4 High-Speed Photography System

The high-speed photo camera employed in the experiment is the Fastcam Ultima APX model produced by Photron company. A high-photosensibility CMOS sensor with one million pixels is used in the camera. CMOS sensor is an advanced type of sensor with good performance of anti-disturbance. An electronic shutter of 4 μs ensures the fidelity and accuracy of images. It is very important for capturing a high-speed transient flame. The camera is composed of high-speed processor, CCD, and LCD, as shown in Fig. 2.3. The maximum speed of the camera is 120,000 fps. The speed can be 2,000 fps with a 1024 × 1024 resolution. Reducing the resolution increases the speed. In the present experiments, the operating speed is chosen to be 10,000–15,000 fps depending on the actual flame speed since hydrogen–air flames generally propagate very fast.

Fig. 2.3 Photograph of high-speed photo camera

2.2.5 Schlieren Optics System

A flame is a strong discontinuity of density, so that there is a large density gradient of flow across the flame front. Because the refractive index of gas is proportional to density, when a light passes through a density discontinuity its position, direction, and path can be deviated. Thus the flame front (density discontinuity) can be captured using optical measurements without disturbing the internal flow field. Schlieren, shadowgraph, and interferometric techniques are three common optical methods for recording changes of flow features [20]. Shadowgraph measures the second derivative of density, namely the displacement of light beam. Schlieren technique records the derivative of density, i.e., the deflection angle of light. Both schlieren and shadowgraph techniques give the variation of density gradient of flow field. Interferometry records the phase difference of light and can be used for quantitative measurement of flow field. In this work, the schlieren technique along with the high-speed camera is used to record the dynamics of flame front propagation.

Schlieren photography has been widely used for many years in science and industry of fluid flow, combustion, boundary layer, gas convection, and wind tunnel. It was proposed by Toepler in 1884 and was applied in the detection of refractive index of optical glass [21]. Schlieren is also commonly used in the visualization of flow field during transient flame propagation [22]. In a schlieren system, the image from the light source is cut by a knife edge, so that the disturbance of lights caused by features of flow field can be described by a light intensity distribution on a plane [23].

The Schlieren system in the present experiment is arranged in a standard Z-configuration with the combustion tube placed in the center of the optical path, as show in Fig. 2.4. The light source is an iodine lamp with a 2.0 mm aperture. The aperture is used to create a point light source. The light rays from the source become parallel light rays after reflected by mirrors M_1 and M_2. After reflected from

Fig. 2.4 Diagram of optical path of the high-speed schlieren system

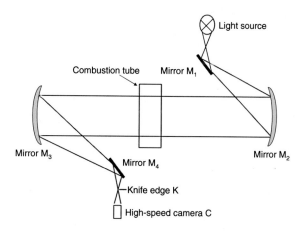

Fig. 2.5 Light source (*upper right*), reflection mirror (*left*), and knife edge (*bottom right*) of the schlieren optics system

M_3 and M_4, the light rays enter into the camera C. The major components of the schlieren system are shown in Fig. 2.5.

2.2.6 Pressure Transducer

The pressure transient inside the duct is measured using a PCB Piezotronics model 112B10 quartz transducer. This transducer is usually used in internal combustion engines. Its main parameters are as follows: (1) response frequency $\geq$ 200 kHz, (2) measuring range $\leq$ 2.0 MPa, and (3) maximum flash temperature 2482 °C.

2.2.7 Data Acquisition Device

The data acquisitor in the experiments is a HIOKI data recorder, model 8826, produced by HIOKI company, as shown in Fig. 2.6. The recorder provides various types of input unit and has 32 channels. The sampling rate of each channel is 1 MS/s.

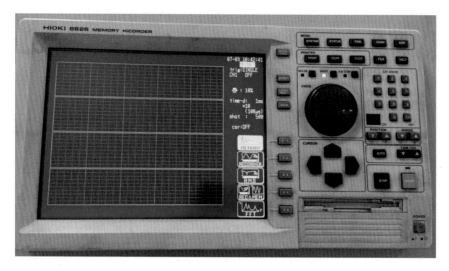

Fig. 2.6 Picture of HIOKI data recorder, model 8826

2.2.8 Synchronization System

The spark igniter, pressure recorder, and high-speed video camera are triggered by the synchronization system in the experiment. The synchronization controller is an OMRON SK20 controller which is a programmable logic controller (PLC). The controller comprises a control panel and a CPU. The commands are programmed into the control panel and stored in the CPU. The triggering time of every terminal is controlled by the CPU. The working principle is shown in Fig. 2.7.

In general, each terminal needs an independent power source with unique voltage, such that each terminal has its own power supply. The direct current (DC) voltages of the high-voltage ignition system and the synchronization controller are 5 and 24 V, respectively. And the DC voltages of both the high-speed camera and data recorder are the same, 3.3 V.

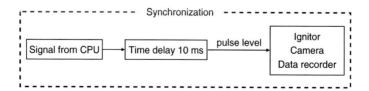

Fig. 2.7 Schematic drawing of working principle of the synchronization system

2.3   Experiment Procedure and Initial Conditions

2.3.1   Methodology

(1) High-speed schlieren photography
    The variations in flame structure, shape, and position as a function of time is
    recorded using the high-speed schlieren system. The flame instabilities that
    develop during flame propagation can be subsequently analyzed.
(2) Pressure measurement
    The pressure dynamics inside ducts is measured using the pressure transducer
    described above. The pressure data together with the high-speed schlieren
    records are used to examine the relationships between flame evolution and
    pressure rise. Then the interactions between flame front and pressure waves
    can be studied. The effects of various parameters, such as equivalence ratio,
    gravity, and opening ratio, on the flame propagation can be scrutinized as well.

2.3.2   Procedure

(1) Setup and test the experimental system, as shown in Fig. 2.1, to ensure that all
    the units of the system are in good state. The vacuum of the combustion duct
    needs to be tested as well before experiment when conducting experiments in
    a closed duct.
(2) Warm up the gas mixing device for 20 min and then fill the duct with
    hydrogen–air mixture using the gas mixing device.
(3) A short time delay of about 30 s is incorporated into the filling sequence
    before ignition, in order to allow the gas mixture to become substantially
    quiescent. This is important to ensure the reproducibility of experiments.
(4) Start the high-voltage ignitor and set the electric voltage to 300 V. And then
    start the data collector, synchronization controller, and high-speed camera.
    After that, start the synchronization system to perform a test.
(5) Store the data recorded by the high-speed schlieren system and pressure
    transducer. Then repeat the experiment.

2.3.3   Initial Parameters

(1) Various compositions of hydrogen–air mixture are used in the experiments.
    The concentration of hydrogen–air mixture used in the experiments ranges
    from 5 to 75 % in steps of 5 %. In addition, the stoichiometric hydrogen–air
    mixture is also used. Therefore, 17 different hydrogen–air mixtures are

employed in the experiments. Each test is repeated three to five times to ensure the reproducibility of the experiments.

(2) The initial pressure and temperature in the duct are 101,325 Pa and 298 K, respectively.
(3) The mixture is at rest before ignition.
(4) The vent should be opened immediately before ignition when performing an experiment in a half-open duct. When conducting an experiment in a closed duct, the vent should be closed during the flame propagation. The vent is opened when the pressure inside the duct exceeds the pressure threshold of the vent.

2.4 Experimental Results and Discussion

Although there have been many experimental, numerical, and theoretical studies on premixed flame propagation in tubes, the flame dynamics has not been sufficiently understood [24–26]. Hydrogen–air flame can be different from that of common hydrocarbon-air flames because of the unique properties of hydrogen, especially the wide flammability limits (equivalence ratio $0.1 \leq \Phi \leq 7.14$, i.e., hydrogen concentration 4.0–75.0 % by volume) and large laminar burning velocity (the maximum can be approximately 3.0 m/s) [27]. In the present work, a series of experiments with a wide range of equivalence ratio are conducted in order to further reveal the behaviors and characteristics of hydrogen–air flames propagating in ducts. The dependence of flame propagation on equivalence ratio, laminar burning velocity, and expansion ratio are examined. The effects of gravity and opening ratio on flame evolution are also discussed. The equivalence ratio is conventionally defined as the ratio of the actual hydrogen–air ratio to the stoichiometric hydrogen–air ratio:

$$\Phi = \frac{n_{\text{fuel}}/n_{\text{air}}}{\left(n_{\text{fuel}}/n_{\text{air}}\right)_{\text{st}}}, \tag{2.2}$$

where n_{fuel} and n_{air} are the mole number of hydrogen and air, respectively. The subscript st denotes stoichiometric.

2.4.1 Hydrogen–Air Flame Propagation in Half-Open Tubes

Under adiabatic conditions a premixed flame can keep a planar front during propagation in a sufficiently narrow channel, whereas in a channel with a width significantly larger than flame thickness it is impossible for a premixed flame to maintain a plane front due to the intrinsic Darrieus–Landau (DL), diffusive-thermal, and

Rayleigh–Taylor (RT) instabilities, etc. [24, 25]. According to Gonzalez et al. [28], the DL instability is always present during the flame formation. On the other hand, Bychkov and Liberman [25] suggested that in order to observe a strongly curved tulip flame as a result of the DL instability in a closed tube, one has to consider tubes of lengths of 200 δ_L and longer. The ducts length in the present work is 530 mm, and the laminar flame thickness of premixed hydrogen–air mixtures is less than 1 mm [29]. So the length scale of the ducts (with length > 530 L_f) in the experiment is large enough to support the DL instability. The diffusive-thermal instabilities play a key role when differential diffusion remains significant in premixed flames of lean hydrogen–air mixtures characterized by a Lewis number less than unity [24]. The RT instability develops at an interface between a heavy and a light matter, when the heavy matter is supported by the light one in a gravitational field [25, 30].

A well-known example of premixed flame propagation in a tube has been suggested and investigated by Clanet and Searby [31] in 1996. Four stages of flame dynamics were proposed and demonstrated [31, 32]: (1) hemispherical expansion flame unaffected by the side walls; (2) finger-shaped flame; (3) elongated flame with the skirt touching the sidewalls; and (4) tulip flame. In the present work, the premixed hydrogen–air flames propagating in the half-open duct exhibit more important information on the premixed flame behaviors and characteristics.

Figure 2.8 shows a series of representative high-speed schlieren images of premixed hydrogen–air flame shape during propagation in the half-open duct at

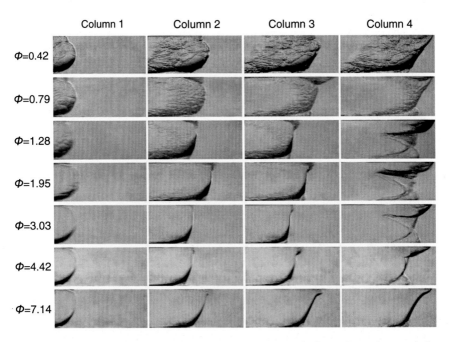

Fig. 2.8 High-speed schlieren images of premixed hydrogen–air flame shape changes during propagation in a half-open duct at various equivalence ratios

various equivalence ratios. The first column indicates the finger-shaped flame front. The second column displays the flame shape with the small parts near the side walls traveling faster than the central region of the flame before the flame front becomes a quasi-plane, as shown in the third column. The fourth column shows the flame shape (tulip flame for some cases) after the formation of quasi-plane flame. Classical tulip flame is observed only at equivalence ratios in the range of $1.17 \leq \Phi \leq 4.05$ in the experiments. The experimental results indicate that at equivalence ratios $0.10 < \Phi < 1.17$ and $4.05 < \Phi \leq 7.14$ the tulip flame could not form, and only the first three stages distinguished by Clanet and Searby can be identified. However, even though no tulip flame forms, a quasi-flatten flame front (actually not exactly plane) will always appear, as shown in Fig. 2.8, except the flame at equivalence ratios very close to the lean flammability limit ($\Phi < 0.36$). In fact, at this low equivalence ratio, the flame advances at a relatively low speed along the upper wall as a result of buoyancy.

In the experiments, an exact plane flame shape is not formed but a quasi-plane flame front is produced at equivalence ratios in the range of $0.36 \leq \Phi \leq 7.14$, as shown in the third column in Fig. 2.8. The quasi-plane flame is an inclined front with the upper and lower small leading tips advancing faster then the central region, forming almost a "T" shape front when $0.67 \leq \Phi \leq 7.14$. The obliquity becomes larger as equivalence ratio gets closer to lean or rich flammability limit. In addition, the upper and/or lower parts of the front have already traveled with a higher speed than the central region just before the formation of the quasi-plane front, as shown in the third column in Fig. 2.8. This phenomenon may be explained by the "squish flow" suggested and discussed by Dunn-Rankin et al. [33] and Gonzalez et al. [28]. "Squish flow" is an accelerated flow which is generated in the unburned region wedged by the flame front and the side walls just ahead of the flame front. The "squish flow" during premixed hydrogen–air flame propagation is more pronounced due to high diffusivity and high laminar burning velocity of hydrogen. "Squish flow" can drive the flame to propagate faster near the walls than in the central region; however, it is not the essential cause that initiates the tulip formation. When equivalence ratio in the range of $0.67 \leq \Phi < 1.17$ and $\Phi > 4.05$, though the squish flow is active, no tulip flame is observed as shown in Fig. 2.8. The "squish flow" is weaker near the lower wall than the upper wall due to the buoyancy. It is also found that with the decreasing of equivalence ratio, the wrinkles become more pronounced. Under the experimental conditions in this work, molecular diffusion acts to stabilize the short wavelength disturbances, and a smooth flame front results for rich hydrogen–air mixtures, whereas the diffusive-thermal instability obviously wrinkles the flame surface in lean hydrogen–air flames.

2.4.2 Hydrogen–Air Flame Propagation in Closed Tubes

Figure 2.9 shows a series of representative high-speed schlieren images of premixed hydrogen–air flame shape changes during propagation at various equivalence ratios

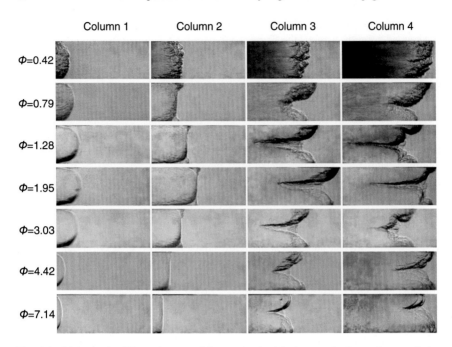

Fig. 2.9 High-speed schlieren images of flame premixed hydrogen–air shape changes during propagation in a closed duct at various equivalence ratios

in closed duct. The first column indicates the ordinary finger-shaped flame front before the flame front transforms significantly into a quasi-plane/plane front, as shown in the second column. The third column designates the complete tulip flame except $\Phi = 0.42$. The fourth column represents the development of tulip flame after its full formation except $\Phi = 0.42$, and the obvious distortion of tulip flame appears at some equivalence ratios. The flame shape shows no distinct change after plane flame formation at $\Phi = 0.42$.

In the experiment, an important finding is that when $\Phi < 0.49$, tulip flame would not be initiated anymore, and when $0.49 \leq \Phi < 0.84$ and $4.22 < \Phi \leq 7.14$, a classical tulip flame, which has been widely described and studied [11, 28, 31, 32, 34], will be produced without apparent distortion. Another outstanding finding in the premixed hydrogen–air propagation is that significant distortion, with the two tulip lips bent/dented obviously from its center to duct sidewalls, will happen to tulip flame after its complete formation when Φ ranges from 0.84 to 4.22, for example $\Phi = 1.28$, 1.95, and 3.03, as shown in Fig. 2.9. Moreover, a normal tulip flame is observed to be reproduced after distortion at $0.84 \leq \Phi \leq 1.12$ and $2.58 \leq \Phi \leq 4.22$. The development of tulip flame after its distortion was not captured at other equivalence ratios as they are out of the measuring range of the schlieren system. The behaviors and dynamics of distorted tulip flame will be discussed in detail in the next two subsections.

In addition, when $0.52 < \Phi < 5.82$, an exact plane flame could never be formed instead the flame displays a quasi-plane front with the leading edges very close to the side walls of the flame front propagating faster than the central region, e.g., at $\Phi = 1.28$, 1.95, 3.03, and 4.22, as shown in the second column in Fig. 2.9. This may also be due to the "squish flow," which has been interpreted in the last section. It is should be noted that a plane flame is formed at other equivalence ratios except extremely low equivalence ratios ($\Phi < 0.36$) because the relatively low laminar burning velocity may reduce the effects of the "squish flow." This proves again that the "squish flow" is not important for tulip formation. And the plane flame front will be wrinkled in very lean hydrogen–air flames due to combustion instabilities, e.g., when $\Phi = 0.42$ as shown in Fig. 2.9.

Experiments in this work also indicate that the flame front instability in closed duct shows the same tendency as that in half-open duct described in the last section, as shown in Fig. 2.9. However, more intense flame disturbance is induced in closed duct due to the higher pressure, and even at extremely high equivalence ratios the wrinkles appear after flame inversion. This can be explained by the fact that as the pressure increases during the propagation in closed duct the flame front becomes thinner, and consequently the hydrodynamic instability is enhanced [24, 35].

2.4.3 Behaviors and Characteristics of Distorted Tulip Flames

The experimental results indicate that more intense tulip flame distortion is generated when equivalence ratio is in the range of $1.12 \leq \Phi \leq 2.58$ than in other cases; meanwhile, the position of the distortion gets closer to the right end of duct, for example $\Phi = 1.28$ and 1.95, as shown in Fig. 2.9. In order to investigate the behaviors and dynamics of the flame with tulip distortion (distorted tulip flame), the premixed hydrogen–air flame propagation accompanied by full tulip distortion development at $\Phi = 3.03$ is taken as the typical case. Figure 2.10 illustrates the typical evolution of premixed hydrogen–air flame with tulip formation and twice distortions at $\Phi = 3.03$ (Fig. 2.10a), as well as schematic classical tulip flame formation (Fig. 2.10b) and schematic premixed hydrogen–air flame shape changes with tulip flame distortion (Fig. 2.10c). In this case, a normal tulip flame is produced once more after its first distortion and then a second distortion is produced again. The flame shapes undergoes more complex changes than classical tulip flame, as shown in Fig. 2.10. The schlieren images show that the tulip is initiated just after the flame burns out at the side walls and then a complete tulip flame is formed. As the flame moves on, two distortions will be produced one after another.

Figure 2.11 shows the relationship between normalized pressure and flame front speed in the propagation direction versus reduced time at $\Phi = 3.03$. The time scale is reduced by W/S_{LO}, where W is the width of the duct cross-section and S_{LO} is the laminar burning velocity. W/S_{LO} is twice of the time scale for the flame to reach the

(a)

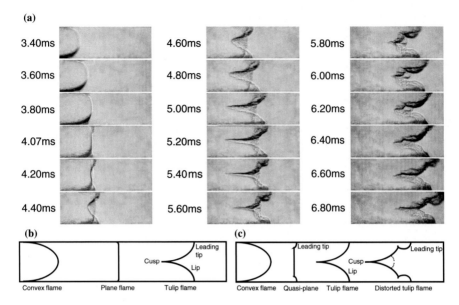

Fig. 2.10 a High-speed schlieren images of premixed hydrogen–air flame at $\Phi = 3.03$.
b Schematic figure showing classical tulip flame formation. **c** Schematic figure showing distorted
tulip flame during premixed hydrogen–air flame propagation

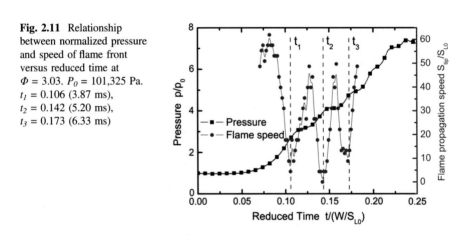

Fig. 2.11 Relationship
between normalized pressure
and speed of flame front
versus reduced time at
$\Phi = 3.03$. $P_0 = 101{,}325$ Pa.
$t_1 = 0.106$ (3.87 ms),
$t_2 = 0.142$ (5.20 ms),
$t_3 = 0.173$ (6.33 ms)

duct side wall. The pressure is reduced by P_0, $P_0 = 101{,}325$ Pa. The flame front
speed is reduced by laminar burning velocities S_{LO}. Before inversion, the flame
speed is determined by the front speed at the symmetry axis, while the speed of the
leading tip close to the upper wall is taken as the front speed after inversion.
Generally, before tulip formation the flame front accelerates quite fast for a short
time with the flame surface area growing exponentially then decelerates sharply due
to the loss of flame surface area [31, 32].

The flame front speed reaches its minimal speed close to zero at $\tau = 0.106$ (3.87 ms) just before the flame is flatten at $\tau = 0.111$ (4.07 ms), as shown in Fig. 2.10a and Fig. 2.11. The first inflection in the pressure trace, which appears at about $\tau = 0.113$ (4.13 ms), correlates very closely with the tulip flame initiation. And then the flame speed increases quickly once more. Just before the first tulip distortion the flame front decelerates fast again. The tulip flame starts to distort at approximately $\tau = 0.137$ (5.00 ms), immediately the flame speed approaches its minimum value of zero at about $\tau = 0.142$ (5.20 ms). And the second inflection of the pressure trace happens just after the first tulip distortion. Thereafter, the flame accelerates with a salient distorted tulip shape. The distorted tulip flame transforms into a normal tulip flame once more about $\tau = 0.169$ (6.20 ms) and last for a quite short time with a much shorter tulip cusp. Another distortion recurs approximately at $\tau = 0.175$ (6.40 ms) near the end of combustion, but with the lips dented more slightly than the first distortion. The flame front speed decreases to a minimum for the third time at $\tau = 0.173$ (6.33 ms), when the pressure trace just begins its third inflection.

On the basis of above analysis, three significant flame shape deformations (one tulip flame, two tulip distortions) were observed as shown in Fig. 2.10. And it can be concluded that the periodic vibrations in the pressure signal correspond to the flame front speed oscillations with 90° phase shift, as shown in Fig. 2.11. When the flame deformation appears accompanied by the flame speed deceleration, the pressure rise comes down due to the loss both of flame surface area and heat of combustion. The flame sharp deceleration is a manifestation of the flame-induced reversal flow analyzed early by Gonzalez et al. [28], and when the flame speed gets close to zero (even equal to zero) it indicates that the absolute velocity value of reversal flow around flame front becomes very close to the laminar burning velocity in propagation direction of flame leading tip. Therefore, the onset of flame deformations coincides with the deceleration both of pressure rise and flame speed.

2.4.4 Comparisons of Distorted Tulip Flame to Classical Tulip Flame

The distortions of tulip flame should be distinguished from the classical tulip disappearance discussed by Starke and Roth [36] and Gonzalez et al. [28], who concluded the propagation of its lateral lips one toward the other finally causes the tulip collapse in tubes of lager aspect ratio. However, in the present study, the tulip flame has a propensity to disappear in this way without apparent tulip distortion, showing classical tulip flame features as schematically shown in Fig. 2.10b, at equivalence ratio in the range of $0.49 \leq \Phi < 0.84$ and $4.22 < \Phi \leq 7.14$. In order to distinguish the tulip distortion from the typical tulip collapse after its full formation, the development of the tulip flame without notable distortion after its full formation is also investigated here for comparison. Figures 2.12 and 2.13 show high-speed

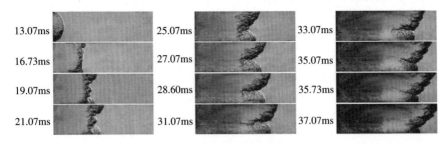

Fig. 2.12 High-speed images of premixed hydrogen–air flame shape changes at $\Phi = 0.59$, representing classical tulip flame in low equivalence ratio range of $0.49 \leq \Phi < 0.84$

schlieren images of the tulip flame initiation, development, and disappearance at $\Phi = 0.59$ and 7.14, representing the classical tulip flame in low equivalence ratio range of $0.49 \leq \Phi < 0.84$ and high equivalence ratio range of $4.22 < \Phi \leq 7.14$, respectively.

It can be seen from Fig. 2.12 that at $\Phi = 0.59$ the tulip flame disappears gradually after its full formation with the upper lateral lip moving at a higher propagation speed toward the lower one, while at $\Phi = 7.14$ the lower lip moves at a higher speed toward the upper one gradually after the complete formation of the tulip flame, and finally the tulip flame vanishes, as shown in Fig. 2.13. The disappearance of tulip flame in this work is consistent with the result obtained by Starke and Roth [8]. From Figs. 2.12 and 2.13, it is known that during the whole process of initiation, development, and disappearance of the classical tulip flame, the tulip distortion does not occur except flame wrinkles which result from various hydrodynamic and combustion instabilities.

The pressure dynamics and flame propagation speed during the flame propagation without noticeable tulip distortion are also examined to further reveal the different characteristics between the distorted tulip flame and the classical tulip flame. Note that the flame propagation speed is defined throughout this thesis is the displacement speed of flame front in the laboratory frame of reference. Figures 2.14a, b illustrate the reduced pressure and reduced flame front speed during the flame propagation of classical tulip flame versus the reduced time at

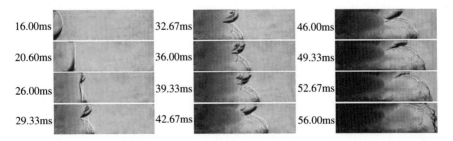

Fig. 2.13 High-speed images of premixed hydrogen–air flame shape changes at $\Phi = 7.14$, representing classical tulip flame in high equivalence ratio range of $4.22 < \Phi \leq 7.14$

Fig. 2.14 Pressure rise and
flame front speed during
classical tulip flame
propagation at $\Phi = 0.59$
(a) and $\Phi = 7.14$ (b)

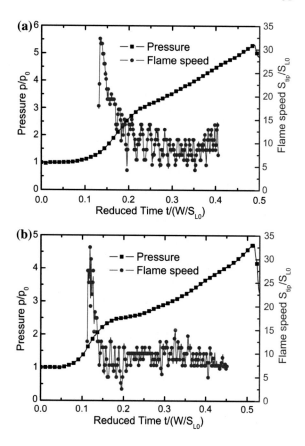

$\Phi = 0.59$ and $\Phi = 7.14$, respectively. Both the pressure traces at the two equivalence
ratios show no apparent vibration before the vent was opened. In fact, it is found
that no obvious vibration, except an inflection around the tulip initiation time, will
happen to the pressure trace at equivalence ratio in the range of $0.49 \leq \Phi < 0.84$ and
$4.22 < \Phi \leq 7.14$. The pressure rise of classical tulip flame during propagation is
similar to that obtained by Dunn-Rankin and Sawyer [11]. The flame front speed at
$\Phi = 0.59$ and $\Phi = 7.14$ also shows different features from that of the distorted tulip
flame. Although the flame speed fluctuates during tulip flame propagation, as
shown in Figs. 2.14a, b, the fluctuation amplitude is much smaller than that of
distorted tulip flame. And the flame speed fluctuation of classical tulip flame is
caused by the vibrations of the flow velocity [8]. These phenomena indicate that
classical tulip flame propagates without sudden shape change and abrupt loss of the
flame surface area. It also implies that the classical tulip flame behaves in a more
stable manner than distorted tulip flame.

Based on the above analysis, the dynamics of distorted tulip flame shows different characteristics from the classical tulip flame. The distorted tulip flame
undergoes more complex shape changes than classical tulip flame. The pressure rise

vibration, correlating very closely with flame deformation, acceleration and deceleration, is a manifestation of more unstable combustion of distorted tulip flame. The relatively smooth pressure rise of classical tulip flame results from more stable combustion.

The "distorted tulip" flame shows more shape changes and behaves in a more unstable manner than the classical tulip flame. Furthermore, it is worth noting that the "distorted tulip" flame is also distinct from the traditional "multi-tulip-shaped" flame [28, 37]. This can be explained by the fact that these "multi-tulip-shaped" flames are initiated by several ignition sources while in the present experiment only one-point ignition source is used. In the case of traditional "multi-tulip-shaped" flames, multiple similar tulip flames propagate at the same time with parallel cusps. However, the cusps of the "distorted tulip" flame are never parallel to each other and the secondary cusps are always smaller than the original one. In addition, Petchenko et al. [38, 39] studied premixed flame propagation from an open end to a closed one in a tube using direct numerical simulation (DNS) and found violent folding and distortion of the flame front due to flame-acoustic interaction. Nevertheless, the flame distortion is different from that in the present work. First, the vessel used in the simulations in [38, 39] is a half-open tube and the flame propagates from the open end to the closed one. Second, the tube width and length considered in the simulations [38, 39] are much smaller than those of the square duct in the present study and any other realistic tube due to the limitation of the length scales for DNS. The inevitably limited computational domain makes the direct comparison of the numerical results to the experiments difficult as indicated in the papers [38, 39]. Finally, the flame distortion can also be distinguished following the shape of the distortion itself.

The flame distortions in [38, 39] appear with two hooked cusps concaved from the sidewall to the center of the duct and a blob of burnt gas pushed into the unburned mixture, which is apparently different from the distortions in the present study. Akkerman et al. [40] investigated flame-flow interaction and generation of vorticity in laminar flame propagation in a closed chamber and found an additional small cusp appearing on the lower lip of a slightly curved flame. The slight distortion in [40] originally occurs near the center of the lip and subsequently corrugates the flame front. Still, this corrugation differs from the present flame distortion since the distortion in the present work is originally initiated close to the flame leading tip and moves toward the primary cusp, and finally develops into a very pronounced secondary cusp in the proximity of the center of the primary tulip lip with the secondary cusp comparable to the primary one (Fig. 2.9). Besides, the ignition source was located at the center of one of the sidewalls in [40].

2.4.5 Effects of Gravity

Gravity can have significant influence on flame stability. The instability is known as gravity instability [24, 25, 41–43]. The upper lip propagates obviously faster than the lower one at low equivalence ratios whereas the later travels at higher speeds than the former due to the buoyancy. As shown in Figs. 2.12 and 2.13, the effects of the gravity become pronounced after tulip inversion. The asymmetrical shape of flame is mainly caused by body force [44]. For the premixed flame propagating in a duct, the body force primarily comes from gravity. The gravity influences the flame front in the following two aspects [45]: (1) the streamlines of unburned mixture close to flame front are deflected by the expanding burnt matter due to the effects of gravity and (2) when the flame is wrinkled, the flame instability can be enhanced or depressed by gravity. This leads to change in burning velocity.

According to Pelce [41], the Froude number and the propagation direction of the flame front may influence the flame shape and flame propagation velocity. The Froude number is defined as

$$\mathrm{Fr} = \frac{gW}{S_L^2}, \tag{2.3}$$

where g is the acceleration of gravity. As in the reference [41], upward propagation is represented conventionally by a negative Froude number and downward propagation by a positive Froude number. Therefore, for the tulip flame the upper lip propagates downward with a positive Froude number and the lower one propagates upward with a negative Froude number in the vertical direction, as shown in Fig. 2.15.

For the lean hydrogen–air mixture characterized by a Lewis number less than unity, the diffusive-thermal effects will play an important role in the flame propagation [24]. In that case, the flame thickness will increase and hence stabilize the flame front for upward propagating flame (Fr < 0) [24, 27, 41] due to gravity. For sufficiently rich hydrogen–air mixture characterized by a Lewis number larger than unity, when a flame propagates upward (Fr < 0), the DL instability is amplified by the RT instability and the flame velocity increases as a result. In the opposite case of a downward propagating flame (Fr > 0), the gravity plays a stabilizing role. The joint effect of the stabilizing gravity and thermal conduction leads to suppression of the flame instability [25, 41]. Based on the above analysis, it can be concluded that

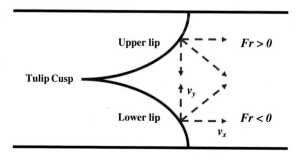

Fig. 2.15 Relationship between Froude number and the propagation of Tulip upper and lower lips

the discrepancies of the propagation speed between the upper lip and the lower one, as shown in Figs. 2.12 and 2.13, are due to the interactions between the flame instabilities and the gravity.

2.4.6   Effects of Equivalence Ratio

According to Clanet and Searby [31], the only relevant parameters influencing the tulip flame formation are the duct cross-section width W, the laminar burning velocity S_L and the nondimensional gas expansion coefficient E as long as flame thickness is small compared to the duct dimensions. The expansion coefficient E is defined as the ratio of the density of unburned mixture to that of the burnt matter. The laminar burning velocity and gas expansion coefficient are directly related to the equivalence ratio Φ. From successful experiment, Dunn-Rankin and Sawyer [11] thought that the formation of a tulip flame is relatively insensitive to equivalence ratio.

Based on the above analysis, the formations of quasi-plane/plane front, tulip flame, and distorted tulip flame depend on the equivalence ratio distinctly both in half-open and closed ducts. And the equivalence ratio range, in which tulip flame can form, is much wider in closed duct ($0.49 \leq \Phi \leq 7.14$) than that in half-open duct ($1.17 \leq \Phi \leq 4.05$). One reason investigated for this circumstance is that the laminar burning velocity increases very quickly due to the increase of the temperature of unburned mixture in the closed duct. Flames propagating in open duct consume fuel in a nearly isobaric regime, since expansion of burning gas in the combustion process may be balanced by the flow of the mixture away from the flame front. However, expansion of burning matter causes adiabatic compression of the fresh fuel ahead of the flame front in closed duct. And the compression of the fuel is accompanied by significant pressure and temperature build up in comparison with the initial values [25]. The dependence of the laminar burning velocity on pressure and temperature is usually expressed as a polynomial function [46]:

$$\frac{S_L}{S_{L0}} = \left(\frac{T}{T_0}\right)^m \cdot \left(\frac{p}{p_0}\right)^n, \tag{2.4}$$

where S_L is the laminar burning velocity at temperature T and pressure p, S_{L0} is the laminar burning velocity at initial temperature T_0 and p_0, m and n are temperature and baric indexes independence of the burning velocity. The flame propagation time in the experiment is rather short and the combustion process can be assumed as adiabatic. According to adiabatic compression law:

$$\frac{T}{T_0} = \left(\frac{p}{p_0}\right)^{(\gamma-1)/\gamma}, \tag{2.5}$$

where $\gamma = c_p/c_v$, is the specific heat ratio, c_p and c_v are specific heats at constant pressure and constant volume correspondingly. During the adiabatic compression of the unburned matter, pressure and temperature change simultaneously. Therefore, the laminar burning velocity can be defined as a function of pressure only [47]:

$$S_L = S_{L0}\left(\frac{p}{p_0}\right)^{\varepsilon},\tag{2.6}$$

where $\varepsilon = m_0 + n_0 - m_0/\gamma$, is the overall thermokinetic index. Values of m, n, and ε can be recommended according to [48]. From Eqs. (2.5) and (2.6), it is known that the laminar burning velocity increases fast due to the adiabatic compression. Therefore, it could be concluded that the adiabatic compression of unburned mixture may plays an important role in causing the wider equivalence ratio range for tulip flame formation in the closed duct.

The time of quasi-plane/plane flame front formation is a critical time, characterizing the significant flame shape change. And this critical time is the flame front inversion time for tulip flame (tulip initiation time). Both the quasi-plane and plane flames are referred to as plane flame here for convenience. The formation time of plane flame versus equivalence ratio in half-open and closed duct is shown in Fig. 2.16. The formation time is reduced by W/S_{L0}. Both in half-open and closed ducts, the formation time as a whole decreases exponentially as the equivalence ratio increases, but the decrease rate is more uniform in half-open duct. The formation time decrease might be a result of the combined action of laminar burning velocity, expansion efficient, and flame instabilities. The time decreases faster in closed duct than that in half-open duct with a larger absolute value of slope when $\Phi < 1.59$, while the opposite is true when $\Phi > 1.59$. The time decreases very slightly with slope close to zero when $\Phi > 2.37$ in closed duct. It also found that between $\Phi = 1.17$ and 4.05, the plane flame formation time in half-open duct is almost the same as that in closed duct within experimental error. And this

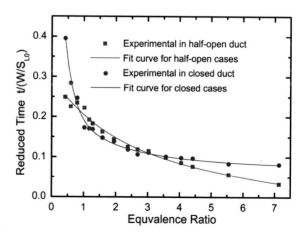

Fig. 2.16 Formation time of plane flame versus equivalence ratio in half-open and closed ducts

equivalence ratio range is just consistent with the tulip formation equivalence ratio range in half-open duct. Therefore, it would be reasonable to conclude that the tulip initiation time in half-open duct is equal to that in closed duct at the same equivalence ratio.

Figure 2.17 shows the relationship between tulip initiation time and first tulip distortion initiation time in closed duct. The time scale is also reduced by W/S_{LO}. The first tulip distortion initiation time and the tulip initiation time nearly share the same tendency versus the equivalence ratio in the distortion range of $0.84 \leq \Phi \leq 4.22$. The inversion time decreases gradually accompanied by slight fluctuation as the equivalence ratio increases. So does the tulip distortion initiation time, except in the range of $0.84 \leq \Phi \leq 1.02$. The discrepancy might be due to the rich shift of the peaking of hydrogen laminar burning velocity.

Figure 2.18 shows the relationship between expansion coefficient and the reduced distances from plane flame and its leading tip to the ignition point in half-open and closed ducts versus equivalence ratio. The distance is reduced by $E \cdot W$. So does that in Fig. 2.19, which will be shown later. The hydrogen adiabatic

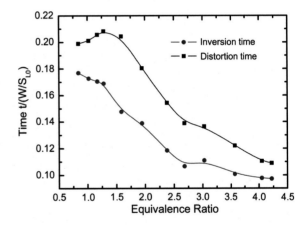

Fig. 2.17 Relationship between the initiation time of tulip flame and the first tulip distortion. The equivalence ratio ranges from 1.02 to 3.57 in closed duct

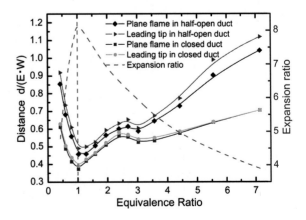

Fig. 2.18 Relationship between expansion coefficient and distances from the plane flame and its leading tip to the ignition point in half-open and closed ducts

Fig. 2.19 Distances from the first distortion initiation position and its leading tip to the ignition point in closed duct compared with those of the corresponding plane flame and leading tip

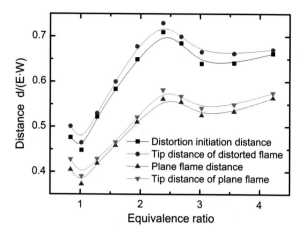

flame temperature can be obtained from an overall energy balance [24, 27], and the expansion coefficient is determined by the ratio of temperature of burnt matter and the temperature of the unburned gas [25, 32]. The position of the plane flame center is taken as the plane position when the plane is a sloping front. The dimensionless distances vary significantly as equivalence ratio increases in half-open duct with a similar trend to those in closed duct. At the beginning, the distances decrease sharply and reach to the minimum value in the vicinity of $\Phi = 1.0$ which is just the location of the maximum of both the hydrogen adiabatic flame temperature and expansion coefficient in air. Indeed, the distance displays an opposite tendency on the whole with the expansion coefficient. Therefore, it is rational to conclude that the formation position of plane flame/initiation position of tulip flame shows an approximately negative correlation with the expansion coefficient versus equivalence ratio, as shown in Fig. 2.18.

Figure 2.19 presents the reduced distances from the first distortion initiation position and its leading tip to the ignition point in closed duct in comparison with those of the corresponding plane flame and leading tip. Both the distances of distortion initiation position and the leading tip of distorted flame keep the same trend with the later located above the former a little with fluctuations. Actually, the first distortion is always initiated near the center of the tulip lips, as shown in Figs. 2.9 and 2.10. The distances of the first distortion/its leading tip also show the same tendency versus equivalence ratio as those of the corresponding plane flame/its leading tip with the former located above the later obviously.

It also can be seen that in half-open duct, the plane flame appears always with leading tips located in front of flame front near the duct side walls, taking the shape of quasi-plane. In closed duct, the distance between the central plane region and the leading tip is less than that in half-open duct, and decreases apparently when $\Phi \geq 5.82$, forming an plane flame subsequently.

2.4.7 Effects of Opening Ratio

In order to examine the influence of opening condition on the flame propagation, an opening ratio σ is defined here as the ratio of open area to the area of the original vent (actual open area/maximum area of vent). Table 2.1 gives the opening ratios and the corresponding diameters of the opening vent at different opening conditions.

Figure 2.20 presents a sequence of high-speed schlieren images of premixed flame propagation for hydrogen–air mixture at opening ratios $\sigma = 0.2$ (a), 0.4 (b), 0.6 (c), 0.8 (d) and 1.0 (e), illustrating typical flame shape changes. Basically, a premixed flame can maintain a plane front in a sufficiently narrow channel under adiabatic conditions, whereas it is impossible for a premixed flame to keep a planar front in a tube with larger width such as that in the present study. A well-known example of premixed flame development in a half-open tube was suggested by Clanet and Searby [31], who proposed four stages of the tulip flame propagation. The experiments for hydrogen–air mixture in the present work show more curious characteristics of premixed flame propagation. The flame evolves differently as the opening ratio increases, as shown in Fig. 2.20. The experimental results show that a noticeable distorted tulip flame is generated after the full formation of a classical tulip flame when the opening ratio is small, namely $\sigma \leq 0.4$. The formation and characteristic features of a distorted tulip flame in the partially open duct (at low opening ratios) are found to be similar to those in a closed duct. Here, take the flame propagation at $\sigma = 0.2$ as a representative example of distorted tulip propagation. After flame inversion at about $t = 6.2$ ms, as shown in Fig. 2.20a, a well-pronounced tulip flame is formed with a slender cusp. The distortions of the tulip flame are initiated close to the tips of the primary tulip tongues. The distortions move backward along the primary tulip tongues and a distorted tulip shape is subsequently created (e.g. flame shape at $t = 7.867$ ms). As the distortions approach the center of the primary tulip tongues, the distorted tulip flame develops into a triple tulip flame (e.g., flame shape at $t = 8.2$ ms). In the meantime, the distortions or secondary cusps appear comparable to the primary one. The distorted tulip flame tends to disappear near the end of the combustion as the primary cusp and the secondary ones propagate to each other, as shown in Fig. 2.20.

When the opening ratio ranges from 0.5 to 1.0, no obvious distorted tulip shape can be observed in the present experiment. As demonstrated above, the interaction between flame front and the combustion generated flow and pressure wave reflected from the right end is responsible for the formation of distorted tulip flame. When the opening ratio is small, the pressure wave can interplay with the flame front and

Table 2.1 Opening ratios and the corresponding diameters of the opening vent at different opening conditions

Opening ratio σ	0.1	0.2	0.3	0.4	0.5	0.6	0.7	0.8	0.9	1.0
Diameter (mm)	12.6	17.9	21.9	25.3	28.3	31.0	33.5	35.8	37.9	40.0

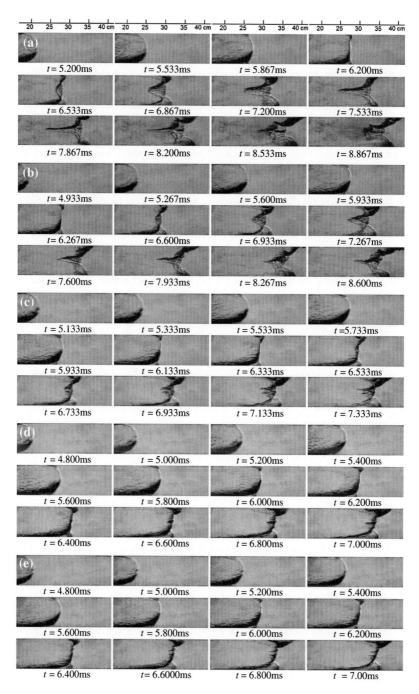

Fig. 2.20 Schlieren images of premixed hydrogen–air flame propagation at opening ratios $\sigma = 0.2$ (**a**), 0.4 (**b**), 0.6 (**c**), 0.8 (**d**), and 1.0 (**e**)

thus drive the flame to display a distorted shape. However, when the opening ratio is large, the effect of the pressure wave may be balanced by the strong flow of the unburned mixture away from the flame front. A tulip flame is produced for all the opening ratios. The larger is the opening ratio, the less pronounced is the tulip flame, as shown in Fig. 2.20. Particularly, when the opening ratio is close to 1.0, the planar front remains in the center of the duct until the end of the propagation process and no remarkable tulip cusp can be formed. On the other hand, with the increase of opening ratio the flame acceleration along the sidewalls becomes stronger. This flame acceleration is caused by the "squish flow," as discussed in Sect. 2.4.1. A squish flow is actually an accelerated flow induced in the unburned region wedged by the flame front and the lateral walls just ahead of the flame. The squish flow can drive the flame near the sidewalls to travel faster than in the central region. The squish flow during premixed hydrogen–air flame propagation is more intensive with the increase of opening ratio since smaller confinement could result in higher flow velocity in the fresh mixture.

Based on the above analysis, it can be concluded that the flame behaves differently with different opening ratio and the flame experiences more drastic flame shape changes as the opening ratio decreases. For this reason, the opening ratio of the vent orifice near the right-end wall of the duct (opposite to the ignition end wall) is an important parameter that affects the flame dynamics.

In the early stages, i.e., the spherical stage and the early phase of the finger stage, the flame in a closed duct propagates nearly at the same speed as those in a half-open tube and exhibits very similar features since the pressure build up in the duct at the early stage can be neglected. The flame propagation (displacement) speed would be smaller in the later stages in a closed duct because additional deceleration is caused by the inability of the gases to expand freely and flow toward the closed end. From this point of view, it could be expected that the flame in the current study travels at the same speed in the early stages for all the opening ratios whereas the flame propagation speed will increase with the increase of opening ratio in the later stages. The main attention of this study is given to the later stages of flame propagation.

Figures 2.21 and 2.22 show the location and propagation speed (displacement speed) of the flame leading tip with time under various opening conditions, respectively, where location is defined as the distance from the ignition point. Before flame inversion, the flame tip along the duct axis is treated as the flame leading tip while the flame tip near the upper sidewall is taken as the leading tip after inversion in this work. Here $H = 4.1$ cm, $S_{L0} = 2.1$ m/s. The increase of opening ratio leads to larger distance at the same time instant, as shown in Fig. 2.21, which indicates that the flame propagation speed increases at the later flame stage with the increase of opening ratio. And the discrepancy becomes more significant with time. Meanwhile, the difference of the flame movement speeds between two adjacent opening ratios approximately becomes smaller as the opening ratio increases.

Oscillations occur in the trajectories of the propagation speed of flame leading tip with location for all the opening conditions, especially for $\sigma \leq 0.5$, as shown in Fig. 2.22. Thermal expansion of combustion products produces movement in the

Fig. 2.21 Location of flame
leading tip as a function of
time at various opening ratios

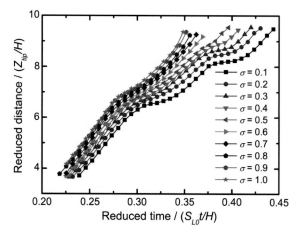

Fig. 2.22 Propagation speed
(displacement speed) of flame
leading tip with location
under different opening
conditions

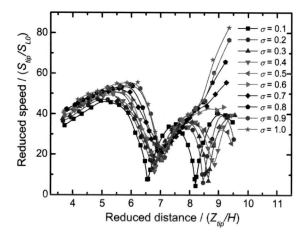

unburned mixture and the confinement of the duct leads the flame to accelerate
quickly. The flame acceleration terminates as the flame skirts reaches the sidewalls
of the duct. As the opening ratio increases the flame accelerates more rapidly, as
shown in Fig. 2.22. Note that since the surface area of the flame front near the
left-end wall is much smaller in comparison with the lateral sides (skirt) of the flame
and the flame approaches the left-end wall at a much lower speed, the contact of the
flame with the left-end wall would further the flame deceleration, but to a lesser
extent. The first flame deceleration stops with flame surface area reduced to a
minimum value as the flame front is flattened out. The formation of tulip flame
results in another increase of the flame surface area, and consequently the flame
accelerates again. When the opening ratio is small, i.e., $\sigma \le 0.5$, the flame accelerates over a few centimeters and then assumes a second flame deceleration, as
shown in Fig. 2.22. The second deceleration is in close connection with the pressure
wave generated by the first contact of the flame with sidewalls, which will be shown

later (Chap. 5). The flame keeps accelerating fast when $\sigma > 0.5$, especially near the orifice. This confirms again that the interaction of flame with pressure wave is pretty weak when $\sigma > 0.094$. Note that although the flame at $\sigma = 0.5$ experiences a second deceleration, it could be too weak to create a remarkably distorted tulip shape.

Figure 2.23 gives the pressure dynamics inside the duct obtained from the pressure transducer at different opening ratios ($P_0 = 101{,}325$ Pa). The pressure at all the opening ratios grow exponentially with the exponential increase of flame surface area at the finger stage. The pressure growth rate drops due to the reduction of combustion products arising from the sudden reduction of flame surface area after the flame skirt touches the sidewalls. This initiates a pressure wave (acoustic wave) which then travels forth and back in the duct. The pressure grows fast again with the generation of the tulip shape. The subsequent oscillations of the pressure dynamics result from the interaction of the flame front with the pressure wave. It is also shown in Fig. 2.23 that increase of the opening ratio leads to a lower pressure growth rate and consequently a smaller peak pressure. The differences of pressure dynamics between the cases at opening ratios close to 1.0 is quite small during the entire combustion process but much larger at opening ratios near 0.1, especially at the later phase of the combustion. Furthermore, the lower is the opening ratio, the larger is the amplitude of pressure oscillations. This implies that the coupling of flame front with pressure wave becomes stronger with decrease of the opening ratio.

As analyzed above, the flame dynamics in the partially open duct is in close connection with the opening ratio at the later stage. The formation of the distorted tulip flame depends on the opening ratio. Figure 2.24 shows the time τ_{wall} at which the flame touches the lateral walls of the duct under various opening conditions. The time τ_{wall} increases gradually as the opening ratio increases. Nevertheless, the increase rate decreases when the opening ratio approaches 1.0. Following the theory by Bychkov et al. [32], the characteristic time t_{wall} when the flame touches the sidewalls can be calculated as:

Fig. 2.23 Pressure dynamics under various opening conditions

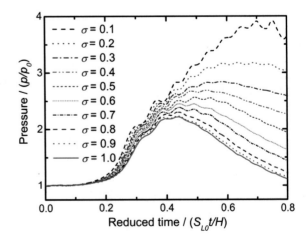

Fig. 2.24 Time when the flame reaches the sidewalls of the duct at various opening ratio

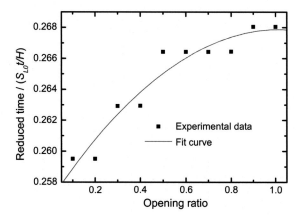

$$t_{wall} = 1/(2\alpha) \cdot H/S_L \cdot \ln[(E+\alpha)/(E-\alpha)], \qquad (2.7)$$

where $\alpha = \sqrt{E \cdot (E-1)}$, $E = 7.22$ is the expansion ratio and S_L is the laminar burning velocity at temperature T and pressure P. When the opening ratio is small, the precompression of the unburned fuel caused by the thermal expansion of the burning matter can be significant. The compression of the fresh mixture is accompanied by considerable pressure and temperature build up compared to the initial values. According to Eq. (2.7), the increase of laminar burning velocity S_L leads to decrease of the time t_{wall} when the flame reaches the sidewalls. Therefore, the dimensionless time τ_{wall} decreases with the decrease of opening ratio since larger confinement results in higher pressure build up (as shown in Fig. 2.23). Figure 2.25 shows the location of the flame leading tip with opening ratio at the time when the flame skirt touches the sidewalls of the duct. The location increases with the increase of opening ratio. This can be explained by the fact that the speed of flame leading tip increases with the opening ratio, as shown in Figs. 2.21 and 2.22.

Fig. 2.25 Flame leading tip location as the flame touches the lateral walls at different opening conditions

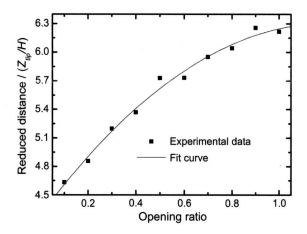

Nevertheless, the increase rate reduces with the increase of opening ratio, as shown in Fig. 2.25.

Figure 2.26 presents the time when the flame front deforms into a planar shape (i.e., flame inversion time) with opening ratio. The flame inversion time increases with the increase of the opening ratio. Following the analytical theory by Bychkov et al. [32], the flame inversion time t_{inv} can be expressed by:

$$t_{inv} = \delta\sigma^{-1}H/S_L, \qquad (2.8)$$

where δ is a model constant comparable to unity. It is clearly shown in Eq. (2.8) that the flame inversion time is inversely proportional to the laminar burning velocity. As remarked above, the decrease of the opening ratio results in an increase of the pressure in the unburned mixture and consequently an increase of the laminar burning velocity (see Eq. (2.5)).

Figure 2.27 shows the location of the planar flame and its leading tip at different opening ratios. Though the flame front at the inversion time is conventionally referred to as plane flame, it is actually not an exact planar shape in the present study because the flame near the sidewalls leads the flame propagation, as shown in Fig. 2.20. In addition, the plane flame is apparently inclined after inversion at high opening ratios. The location of the plane flame is defined here as the location of the flame front at the duct axis while the leading tip is the flame tip close to the upper sidewall. The location of the plane flame undergoes an exponential increase until σ reaches to 0.8. The location of the plane flame increases very slowly when $\sigma > 0.8$. The location of the leading tip of the plane flame shows a quite similar trend, but at a higher increase rate. The discrepancy between the locations of the plane flame and its leading tip grows larger as the opening ratio increases. This is thought to be due to two reasons. First, a more inclined plane flame forms as the opening ratio increases, as shown in Fig. 2.20. Second, the flame front near the sidewalls travels faster with the increase of opening ratio due to the enhanced squish flow by the higher flame propagation speed (see Fig. 2.22). As aforementioned, when the

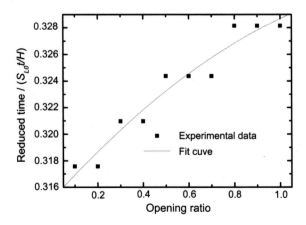

Fig. 2.26 Flame inversion time as a function of opening ratio

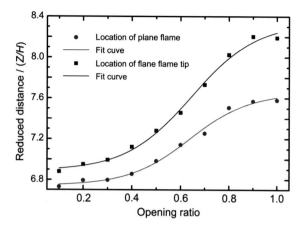

Fig. 2.27 Location of plane flame and its leading tip with opening ratio

opening ratio approaches 1.0 the impact of the opening ratio is not significant any more.

Figure 2.28 shows the onset time of the distorted tulip flame and the location of its leading tip. The initiation time of the tulip distortion increases as the opening ratio increases in the range of $\sigma \leq 0.4$. The formation of a distorted tulip flame has a close connection with the pressure wave and a tulip distortion is created immediately after the pressure wave passes the flame front from the unburned side (will be shown in Chaps. 4 and 5). The compression of the unburned mixture leads to an increase of the temperature. The sound speed c_s is increased as a result as follows:

$$c_s = \sqrt{\gamma R_g T}, \tag{2.9}$$

where R_g is the gas constant of the mixture. Therefore, the time required for the formation of tulip distortion reduces with the increase of the sound speed. The location of the flame leading tip at the onset time of tulip distortion increases almost linearly with the increase of opening ratio due to the increase of the propagation speed of the flame leading tip.

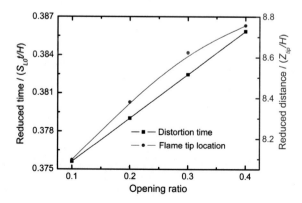

Fig. 2.28 Initiation time of tulip flame distortion and the corresponding flame leading tip location

In summary, the opening ratio has a significant influence on the combustion dynamics, including the flame behaviors and pressure dynamics. Smaller opening ratio leads to more drastic flame shape changes. When the opening ratio is in the range of $\sigma \leq 0.4$, a noticeably distorted tulip flame can be produced after the full formation of a classical tulip flame. When $\sigma > 0.4$, no remarkable tulip distortion can be observed. The propagation speed of flame leading tip increases with the increase of the opening ratio. The flame leading tip undergoes a second strong deceleration with $\sigma \leq 0.5$. The flame shape and propagation speed of the flame leading tip are in close connection with the pressure wave, especially for combustion processes at the low opening ratio. Both the growth rate and oscillation amplitude of the pressure inside the duct increase as the opening ratio decreases. The flame tip locations with the initiation of tulip and distorted tulip flames increase with the increase of the opening ratio. The effect of the opening ratio on the flame and pressure dynamics becomes weak when it is close to 1.0.

2.5 Summary

(1) Premixed hydrogen–air flames propagating in a duct exhibit consistent deformations with other typical gas fuels, forming classical tulip flames, only in special equivalence ratio ranges, $1.17 \leq \Phi \leq 4.05$ in the half-open duct, and $0.49 \leq \Phi < 0.84$ and $4.22 < \Phi \leq 7.14$ in the closed duct, respectively.

(2) It is found that outstanding tulip distortions are generated after full tulip formation in the equivalence ratio range of $0.84 \leq \Phi \leq 4.22$ in closed duct. This curious flame phenomenon is referred to as distorted tulip flame. In same cases, the classical tulip flame is observed to be produced again after the first distortion and then a second distortion will appear subsequently. The onset of flame deformations coincides with the deceleration both of pressure rise and flame front speed. The behaviors of distorted tulip flame are distinguished from those of classical tulip flame in detail. The classical tulip flame develops, and then disappears gradually without distortion after its full formation. The dynamics of distorted tulip flame is different from that of classical tulip flame. The distorted tulip flame undergoes more complex shape changes and more unstable combustion process than the classical tulip flame.

(3) The initiation of flame shape changes, including the formation of quasi-plane/plane flame, tulip flame, and tulip distortion, greatly depend on the mixture composition. The equivalence ratio range that can form tulip flame is much wider in the closed duct ($0.49 \leq \Phi \leq 7.14$) than that in the half-open duct ($1.17 \leq \Phi \leq 4.05$). The dimensionless time of formation quasi-plane/plane flame and initiation of tulip flame as a whole exponentially decreases as the equivalence ratio increases both in the half-open and closed ducts. And the tulip flame formation time in half-open duct is nearly equal to that in closed duct at the same equivalence ratio. The dimensionless distances from quasi-plane/plane flame to ignition point vary significantly as

equivalence ratio increases in half-open duct with the same trend as those in closed duct. The formation position of the quasi-plane/plane shows an approximately negative correlation with the expansion coefficient. The distances of the first tulip distortion/its leading tip also show the same tendency versus equivalence ratio as those of plane flame/its leading tip.

(4) The "squish flow" is active and can cause local deformations of flame shape near the duct walls. However, it may be unimportant for tulip flame formation. The gravity has a noticeable influence on the tulip flame and can make the tulip flame collapse in different way between low and high equivalence ratios, but may not lead to substantial flame shape change.

(5) Both gravity and opening ratio have significant influence on the flame dynamics. The upper lip propagates obviously faster than the lower one at low equivalence ratios whereas the later travels at higher speeds than the former due to the buoyancy. Smaller opening ratio leads to more drastic flame shape changes. When the opening ratio is in the range of $\sigma \leq 0.4$, a noticeably distorted tulip flame can be produced after the full formation of a classical tulip flame. The propagation speed of flame leading tip increases with the increase of the opening ratio. The flame leading tip undergoes a second strong deceleration with $\sigma \leq 0.5$. Both the growth rate and oscillation amplitude of the pressure inside the duct increase as the opening ratio decreases. The flame tip locations with the initiation of tulip and distorted tulip flames increase with the increase of the opening ratio.

References

1. Hirano T (2002) Combustion science for safety. Proc Combust Inst 29:167–180
2. Ng HD, Lee JHS (2008) Comments on explosion problems for hydrogen safety. J Loss Prev Process Ind 21:136–146
3. Ciccarelli G, Dorofeev S (2008) Flame acceleration and transition to detonation in ducts. Prog Energy Combust Sci 34:499–550
4. Zhou K, Li Z (2000) Flame front acceleration of prone-air deflagration in straight tubes. Explosion Shock Waves 20:137–142 (in Chinese)
5. Yu J, Zhou C, Liu R, Yan Q (2004) Experimental study on propagating characteristics of flame and pressure waves in explosion of combustible gases. Nat Gas Ind 24:87–90 (in Chinese)
6. Chen X, Sun J, Liu Y, Liu X, Chen S, Lu S (2006) Fine structure of premixed propane/air flame during transition from laminar to turbulent regime. Chin Sci Bull 51:2920–2925 (in Chinese)
7. Fairweather M, Hargrave GK, Ibrahim SS, Walker DG (1999) Studies of premixed flame propagation in explosion tubes. Combust Flame 116:504–518
8. Starke R, Roth P (1986) An experimental investigation of flame behavior during cylindrical vessel explosions. Combust Flame 66:249–259
9. Starke R, Roth P (1989) An experimental investigation of flame behavior during explosions in cylindrical enclosures with obstacles. Combust Flame 75:111–121
10. Dunn-Rankin D, Sawyer RE (1985) Interaction of a laminar flame with its self-generated flow during constant volume combustion. In: Proceedings of the 10th ICDERS, Berkley, California, F August

11. Dunn-Rankin D, Sawyer RF (1998) Tulip flames: changes in shape of premixed flames propagating in closed tubes. Exp Fluids 24:130–140
12. Sato Y, Iwabuchi H, Groethe M, Merilo E, Chiba S (2006) Experiments on hydrogen deflagration. J Power Sources 159:144–148
13. Molkov VV, Makarov DV, Schneider H (2007) Hydrogen-air deflagrations in open atmosphere: large eddy simulation analysis of experimental data. Int J Hydrogen Energy 32:2198–2205
14. Baraldi D, Kotchourko A, Lelyakin A, Yanez J, Gavrikov A, Efimenko A, Verbecke F, Makarov D, Molkov V, Teodorczyk A (2010) An inter-comparison exercise on CFD model capabilities to simulate hydrogen deflagrations with pressure relief vents. Int J Hydrogen Energy 35:12381–12390
15. Baraldi D, Kotchourko A, Lelyakin A, Yanez J, Middha P, Hansen OR, Gavrikov A, Efimenko A, Verbecke F, Makarov D, Molkov V (2009) An inter-comparison exercise on CFD model capabilities to simulate hydrogen deflagrations in a tunnel. Int J Hydrogen Energy 34:7862–7872
16. Wen JX, Madhav RVC, Tam VHY (2010) Numerical study of hydrogen explosions in a refuelling environment and in a model storage room. Int J Hydrogen Energy 35:385–394
17. Groethe M, Merilo E, Sato Y (2005) Large-scale hydrogen deflagrations and detonations. In: International conference on hydrogen safety. Pisa, Italy
18. Groethe M (2002) Hydrogen deflagration safety studies in a semi-open space. In: 14th world hydrogen energy conference. Montreal, QC, Canada
19. Zhao H (1996) Principles of gas and dust explosions. Press of Beijing University of Technology, Beijing (in Chinese)
20. Kleine H, Timofeev E, Takayama K (2005) Laboratory-scale blast wave phenomena-optical diagnostics and applications. Shock Waves 14:343–357
21. Feng T, Liu C, Zhao R, Wang F (1994) Schlieren methods: a review of techniques. J Ballist 2:89–96 (in Chinese)
22. He X, Yang Y, Wang E, Liu Z (2004) Effects of obstacle on premixed flame microstructure and flame propagation in methane/air explosion. J China Coal Soc 29:186–189 (in Chinese)
23. Liao G, Wang X, Qin J (2003) Experimental diagnostics of thermal hazards. University of Science and Technology of China Press, Hefei (in Chinese)
24. Matalon M (2009) Flame dynamics. Proc Combust Inst 32:57–82
25. Bychkov VV, Liberman MA (2000) Dynamics and stability of premixed flames. Phys Rep 325:116–237
26. Xiao HH, Wang QS, He XC, Sun JH, Yao LY (2010) Experimental and numerical study on premixed hydrogen/air flame propagation in a horizontal rectangular closed duct. Int J Hydrogen Energy 35:1367–1376
27. Law CK (2006) Combustion physics. Cambridge University Press, New York
28. Gonzalez M, Borghi R, Saouab A (1992) Interaction of a flame front with its self-generated flow in an enclosure—the tulip flame phenomenon. Combust Flame 88:201–220
29. Aung KT, Hassan MI, Faeth GM (1997) Flame stretch interactions of laminar premixed hydrogen/air flames at normal temperature and pressure. Combust Flame 109:1–24
30. Landau LD, Lifshitz EM (1987) Fluid mechanics. Pergamon Press, Oxford
31. Clanet C, Searby G (1996) On the "tulip flame" phenomenon. Combust Flame 105:225–238
32. Bychkov V, Akkerman V, Fru G, Petchenko A, Eriksson LE (2007) Flame acceleration in the early stages of burning in tubes. Combust Flame 150:263–276
33. Dunn-Rankin D, Barr PK, Sawyer RF (1986) Numerical and experimental study of "tulip" flame formation in a closed vessel. Proc Combust Inst 21:1291–1301
34. Matalon M, Metzener P (1997) The propagation of premixed flames in closed tubes. J Fluid Mech 336:331–350
35. Law CK (2006) Propagation, structure, and limit phenomena of laminar flames at elevated pressures. Combust Sci Technol 178:335–360
36. Roth R, Starke P (1986) An experimental investigation of flame behavior during cylindrical vessel explosions. Combust Flame 66:249–259

37. Pocheau A, Kwon CW (1989) Proceedings of A.R.C. Colloquium. C.N.R.S.-P.I.R.S.E.M., Paris, p 62
38. Petchenko A, Bychkov V, Akkerman V, Eriksson LE (2006) Violent folding of a flame front in a flame-acoustic resonance. Phys Rev Lett 97:164501
39. Petchenko A, Bychkov V, Akkerman V, Eriksson LE (2007) Flame-sound interaction in tubes with nonslip walls. Combust Flame 149:418–434
40. Akkerman VB, Bychkov VV, Bastiaans RJM, de Goey LPH, van Oijen JA, Eriksson LE (2008) Flow-flame interaction in a closed chamber. Phys Fluids 20:055107
41. Pelce P (1985) Effect of gravity on the propagation of flames in tubes. J Phys 46:503–510
42. Khokhlov AM, Oran ES, Wheeler JC (1996) Scaling in buoyancy-driven turbulent premixed flames. Combust Flame 105:28–30
43. Libby PA (1989) Theoretical analysis of the effect of gravity on premixed turbulent flames. Combust Sci Technol 68:15–33
44. Molkov V, Dobashi R, Suzuki M, Hirano T (2000) Venting of deflagrations: hydrocarbon-air and hydrogen-air systems. J Loss Prev Process Ind 13:397–409
45. Wang Y, Lei Y, Zhang X, Konig J, Eigenbrod C (2002) Buoyancy influence on wrinkled premixed V-flames. J Combus Sci Technol 8:493–497 (in Chinese)
46. Poinsot T, Veynante D (2005) Theoretical and numerical combustion, 2nd edn. Edwards RT Inc, Philadelphia
47. Nagy J, Conn J, Verakis H (1969) Explosion development in a spherical vessel. Technical reports RI 7279, US Department of Interior, Bureau of Mines
48. European NoE HySafe (2007) http://www.hysafe.org/BRHS. Biennial report on hydrogen safety (version 1.2)

Chapter 3
Numerical Simulations of Dynamics of Premixed Hydrogen-Air Flames Propagating in Ducts

3.1 Introduction

Computational Fluid Dynamics (CFD) is more and more widely used in combustion and explosion science and engineering with the fast development of computer technology [1]. CFD is based on numerical computation and graphical visualization techniques and can be employed to analyze various physical and chemical phenomena involving fluid dynamics, chemical reaction, and heat and mass transfer [2]. CFD together with conventional experimental measurement and theoretical analysis constitute a "three-dimensional" research system of fluid dynamics, heat transfer and combustion, as shown in Fig. 3.1 [3].

The basic idea of CFD is to replace the continuous problem domain with a discrete domain using a mesh. The conservation equations for mass, momentum, and energy are solved throughout the discretized domain to obtain the approximate solution of the problem [4, 5]. In other word, CFD is a numerical simulation of fluid flow by solving the governing equations, i.e., conservation equations for mass, momentum, energy, and species.

Compared to experiment, CFD has many advantages. First, CFD can help to save plenty of time and cost since it can give results without design and construction of experimental setup which is usually expensive and time-consuming. Second, CFD can be used to investigate large-scale problems with low cost. This is especially important for explosion research because it is very expensive, difficult, and dangerous to conduct a large-scale experiment of explosion. Theoretical analysis often requires simplifying a problem, although it can yield generic results. For most practical issues, it may be impossible to obtain an analytic solution. In contrast, CFD can provide numerical solution for a broad range of physical phenomena. Experiment is the basis of development of both theories and numerical methods. Generally, the reliability and accuracy of a CFD tool should be validated against experiments. CFD also has disadvantages since it only provides an approximation instead of an exact solution. Moreover, a CFD method/algorithm

© Springer-Verlag Berlin Heidelberg 2016
H. Xiao, *Experimental and Numerical Study of Dynamics of Premixed Hydrogen-Air Flames Propagating in Ducts*,
Springer Theses, DOI 10.1007/978-3-662-48379-4_3

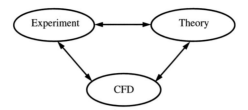

Fig. 3.1 Relationships between CFD, experiment and theory of fluid dynamics

needs to be verified against mathematical model before it can be adopted in an actual calculation. In particular, it is prohibitively expensive to directly resolve a turbulent flow due to the limits of current computer performance. Turbulence modeling is a usual way to numerically solve a turbulent flow. Nevertheless, the modeling of interactions between turbulent flow and combustion remains an unsolved scientific problem.

Many turbulent models have been extensively employed in science and engineering because they are computationally much cheaper than direct numerical simulation (DNS). There are six types of numerical approach of turbulent flow which are commonly used [6]: zero-equation model, one-equation model, two-equation model, Reynolds stress equation model, large eddy simulation model, and DNS. The first four approaches are referred to as Reynolds averaged Navier–Stokes (RANS) models. Three common turbulence approaches are introduced as follows:

(1) k-ε model. It is a typical two-equation model. k and ε are turbulent kinetic energy and turbulent dissipation rate, respectively. This model is the most widely used model in engineering. It comprises standard k-ε model, RNG k-ε model, and realizable k-ε model [2].

(2) Large eddy simulation (LES). The transport of momentum, mass, energy, and other physical quantities in a turbulent system mainly depends on the large eddies. Large eddies are closely connected to the problem considered. They are determined by boundary conditions and independent of each other [2]. The assumption leading to LES is that small dissipative eddies are isotropic and homogeneous. In LES, large-scale turbulence (large eddies) are calculated by resolving Navier–Stokes equations [7], while the smallest turbulence (small eddies) are taken into account using sub-grid scale model (SGS model). SGS model is the key issue of LES. There are various SGS models, such as Smagorinsky model and RNG models. LES is a method between RANS and DNS. The computation cost of LES is related to grid resolution. In general, LES is more expensive than RANS, but much cheaper than DNS.

(3) Direct numerical simulation (DNS). DNS resolves every scale of the turbulent flow and thus requires extremely fine grid. Meanwhile, the time step should also be small enough to resolve the detailed turbulent structures which change drastically over space and time. DNS does not include any turbulence model and can theoretically yield accurate result [8]. Nonetheless, the computational

cost of DNS is very high, so that at present it is computationally prohibited to conduct DNS in computations of any practical engineering systems and DNS can only be used in small-scale flow at low Reynolds number [9, 10]. With respect to DNS in premixed flame propagation in tube, an example can be found in the work by Bychkov et al. [11] who conducted a DNS study on laminar flame acceleration in small-scale long open tubes for the validation of an analytical model.

As remarked above, turbulence can be important for flame acceleration and DDT in explosions. Numerous CFD codes and softwares have been developed for simulating turbulent combustion. However, turbulent combustion generally occurs at scales blow the grid resolution, namely at sub-grid scale. Thus the combustion needs to be modeled. The combustion model for LES is different from that for RANS. Nonetheless, RANS combustion models can be modified for application in LES. Most of these combustion models are based on average flow field. Three common combustion models are: (1) flame surface density (FSD) model, which a geometrical method, (2) probability density function (PDF) model, which is a statistical approach, (3) turbulent mixing model. The LES combustion models include G-equation, FSD, thickened flame model, and progress variable [12]. Eddy break-up model is usually used in RANS calculations.

CFD methods and models should be verified and validated before they can be used in practical systems. For gas explosion CFD codes, the data used for validation can be information of flame speed, overpressure, and flame front dynamics [13–20]. At present, there are various validated CFD tools/softwares for numerical simulation of combustion and explosion, such as FLACS, AutoReaGas, ANSYS-FLUENT, and OpenFOAM. It should be noted that CFD is a developing technology and there may be no such a code/software that can reproduce all the properties of flow, combustion, flame instabilities, DDT, and detonation under different conditions.

3.2 Models and Methods

3.2.1 Physical Model

Premixed flame propagation in a tube is substantially an unsteady reacting flow process. The reaction zone is very thin and separates burnt zone from unburned zone. The cold fresh mixture is consumed as the flame front moves into the unburned zone and turned to hot combustion products. The flame propagation process can be simplified by a physical model with three zones, i.e., burnt, reaction, and unburned zones, as shown in Fig. 3.2. In this model, the flame front is actually a strong moving discontinuity with chemical reaction. The flame propagation is physically a 3D

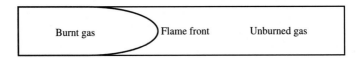

Fig. 3.2 Schematic showing the physical model of premixed flame propagating in a tube

reacting flow phenomenon. The flow and combustion processes can be modeled when the thermal-physical and chemical properties are known. In some cases, the flame propagation can be further simplified into a 2D reacting flow phenomenon. During the combustion process, the flow regime may change from laminar to turbulent. The process becomes more complicated when turbulence occurs.

3.2.2 Mathematical Model and Governing Equations

Weak ignition of a quiescent reactive mixture in a tube generally initiates a laminar flame. The flame then can be wrinkled due to hydrodynamic and combustion instabilities as it continues to propagate. The wrinkled flame may be further disturbed by shear flow near the walls in a sufficiently long tube. Transition from laminar to turbulent combustion takes place as turbulence grows in the unburned mixture. In the present study, the combustion vessels considered are smooth ducts with moderate length scale. Turbulence may be generated during the entire burning process in such a case [21, 22]. In addition, self-turbulization could occur to a premixed hydrogen-air flame. Two types of flow models are employed here to simulate the flame propagation. First, in order to examine the flame characteristics and mechanisms in the absence of turbulence, the flame is simulated as a laminar reactive flow. Then, a LES approach which can handle laminar, transitional, and turbulent flows is used to calculate the flame propagation process. It is hoped that the mechanisms of flame propagation and combustion regimes during the burning process in a duct can be elucidated. The numerical simulations will be compared with experimental results to examine the accuracy of the numerical methods.

3.2.2.1 Laminar Model

The transient premixed flame propagation in a laboratory-scale tube can be modeled as a 2D or 3D laminar reacting flow. The governing equations consist of conservation equations of mass, momentum, energy, and species:

$$\frac{\partial \rho}{\partial t} + \frac{\partial}{\partial x_i}(\rho u_i) = 0, \tag{3.1}$$

$$\frac{\partial}{\partial t}(\rho u_i) + \frac{\partial}{\partial x_j}(\rho u_i u_j) = -\frac{\partial p}{\partial x_i} + \frac{\partial \sigma_{ij}}{\partial x_j}, \tag{3.2}$$

where σ_{ij} is the stress tensor: $\sigma_{ij} = \mu\left(\frac{\partial u_i}{\partial x_j} + \frac{\partial u_j}{\partial x_i}\right) - \frac{2}{3}\mu\frac{\partial u_i}{\partial x_i}\delta_{ij}$.

$$\frac{\partial}{\partial t}(\rho e) + \frac{\partial}{\partial x_i}(u_i(\rho e + p)) = \frac{\partial}{\partial x_i}\left(k\frac{\partial T}{\partial x_i} - \sum h_m J_m + u_j \sigma_{ij}\right) + \dot{Q}_c, \tag{3.3}$$

$$\frac{\partial}{\partial t}(\rho Y_m) + \frac{\partial}{\partial x_i}(\rho u_i Y_m) = \frac{\partial}{\partial x_i}\left(\rho D_m \frac{\partial Y_m}{\partial x_i}\right) + \dot{\omega}_m, \tag{3.4}$$

where ρ is the density, t is the time, u_i is the velocity component, p is the pressure, μ is the viscosity, k is the heat conduction coefficient, T is the temperature, $e = h - p/\rho + u_i^2/2$ is the specific internal energy, and $\dot{Q}_c$ is the heat source term due to the chemical reactions. Y_m, D_m, h_m, J_m, and $\dot{\omega}_m$ are the mass fraction, diffusivity coefficient, specific enthalpy, diffusion flux, and reaction rate of species m, respectively. The chemical reactions are taken into account by using a thickened flame model which will be described below (Sect. 3.2.3).

The gases are assumed to behave as ideal gas and the equation of state of an ideal gas is added:

$$p = \rho R T, \tag{3.5}$$

where R is the gas constant of mixture.

3.2.2.2 Large Eddy Simulation (LES) Model

The objective of LES here is to provide more accurate numerical simulations by accounting for more physical mechanisms, such as flame wrinkling due to instabilities and turbulence in the incoming flow. The governing equations, employed here for the LES, are obtained by filtering three-dimensional (3D) instantaneous conservation equations for mass, momentum, and energy for compressible Newtonian fluid that can be found elsewhere [14, 23]:

$$\frac{\partial \overline{\rho}}{\partial t} + \frac{\partial}{\partial x_j}(\overline{\rho}\tilde{u}_j) = 0, \tag{3.6}$$

$$\frac{\partial \overline{\rho}\tilde{u}_i}{\partial t} + \frac{\partial}{\partial x_j}(\overline{\rho}\tilde{u}_j\tilde{u}_i) = -\frac{\partial \overline{p}}{\partial x_i} + \frac{\partial}{\partial x_j}\left(\mu_{\text{eff}}\left(\frac{\partial \tilde{u}_i}{\partial x_j} + \frac{\partial \tilde{u}_j}{\partial x_i} - \frac{2}{3}\frac{\partial \tilde{u}_k}{\partial x_k}\delta_{ij}\right)\right) + \overline{\rho}g_i \tag{3.7}$$

$$\frac{\partial}{\partial t}\left(\overline{\rho}\,\tilde{e}\right) + \frac{\partial}{\partial x_j}\left(\tilde{u}_j(\overline{\rho}\tilde{e} + \overline{p})\right) =$$

$$\frac{\partial}{\partial x_j}\left(\frac{\mu_{\mathrm{eff}}c_p}{\mathrm{Pr}_{\mathrm{eff}}}\frac{\partial \tilde{T}}{\partial x_j} - \sum_m \tilde{h}_m\left(-\frac{\mu_{\mathrm{eff}}}{\mathrm{Sc}_{\mathrm{eff}}}\frac{\partial \tilde{Y}_m}{\partial x_j}\right) + \tilde{u}_i\mu_{\mathrm{eff}}\left(\frac{\partial \tilde{u}_i}{\partial x_j} + \frac{\partial \tilde{u}_j}{\partial x_i} - \frac{2}{3}\frac{\partial \tilde{u}_k}{\partial x_k}\delta_{ij}\right)\right) + \overline{S}_e.$$

$$(3.8)$$

LES filtered (over bar) and mass-weighted (Favre) filtered (tilde) quantities are introduced correspondingly as [24]:

$$\overline{\phi}(x,t) = \int_V \phi\,(x',t)\,G(x,x')\,d^3x', \qquad (3.9)$$

$$\overline{\rho}(x,t)\,\tilde{\phi}(x,t) = \int_V \rho(x',t)\,\phi(x',t)\,G(x,x')\,d^3x'. \qquad (3.10)$$

Here the LES filter in a physical space is defined as $G(x,x') = 1/V_{CV}$ if $x' \in V_{CV}$ and $G(x,x') = 0$ elsewhere (V_{CV} is the volume of the computational cell) and it is implicitly introduced by finite volume discretization. The source term for the energy equation is associated with the chemical reaction rate $S_e = H_c \cdot S_c$ and will be considered in the combustion model section below.

Effective viscosity is calculated according to the renormalization group (RNG) theory [25]. The RNG method allows one to evaluate turbulent transport coefficients and turbulence transport equations for the large-scale (slow) modes, and, as claimed, "doesn't include any experimentally adjustable coefficients" [25]. The approach is capable of modeling fluid flow within the limits of both laminar and high Reynolds number flow regimes:

$$\mu_{\mathrm{eff}} = \mu\left[1 + H\left(\frac{\mu_s^2\,\mu_{\mathrm{eff}}}{\mu^3} - 100\right)\right]^{1/3}, \qquad (3.11)$$

where $\mu_s = \overline{\rho}\left(0.157\,V_{CV}^{1/3}\right)^2\sqrt{2\tilde{S}_{ij}\tilde{S}_{ij}}$, $H(x)$ is the Heaviside function. The modeling expression for SGS velocity is:

$$u' = \sqrt{\frac{2}{3}}\cdot\frac{\mu_t}{\rho\cdot L_{\mathrm{SGS}}}. \qquad (3.12)$$

In this study the molecular Prandtl number is set to $\mathrm{Pr} = 0.7$. The effective Prandtl number $\mathrm{Pr}_{\mathrm{eff}}$ derived from the RNG theory [26] for nonreactive flows is accepted for calculations:

$$\left|\frac{1/\mathrm{Pr_{eff}} - 1.3929}{1/\mathrm{Pr} - 1.3929}\right|^{0.6321} \left|\frac{1/\mathrm{Pr_{eff}} + 2.3929}{1/\mathrm{Pr} + 2.3929}\right|^{0.3679} = \frac{\mu}{\mu_{\mathrm{eff}}}. \tag{3.13}$$

The same is applied for the laminar and turbulent Schmidt numbers.

The species transport equation for the premixed combustion system is recast in the form of the progress variable equation [14, 23]:

$$\frac{\partial}{\partial t}(\bar{\rho}\tilde{c}) + \frac{\partial}{\partial x_j}(\bar{\rho}\tilde{u}_j\tilde{c}) = \frac{\partial}{\partial x_j}\left(\frac{\mu_{\mathrm{eff}}}{Sc_{\mathrm{eff}}}\frac{\partial\tilde{c}}{\partial x_j}\right) + \bar{S}_c. \tag{3.14}$$

The mass burning rate during combustion is described according to the gradient method [14, 23]:

$$\bar{S}_c = \rho_u S_t \left|\mathrm{grad}\,\tilde{c}\right|, \tag{3.15}$$

where $\left|\mathrm{grad}\,\tilde{c}\right|$ is the gradient of the progress variable. The gradient method ensures that the prescribed mass burning rate $\rho_u S_t$ takes place in the simulations independent of the grid resolution and the numerical flame thickness, which typically spreads over 3–5 control volumes. In fact, though Equation (3.14) allows the flame front to diffuse indefinitely, it is not prejudicial to the LES performed because the time of the flame propagation is very short compared to the time of the diffusion of the flame front. It is unaffordable to resolve the 3D transient flame with a detailed chemistry scheme at a scale of the experiment carried out and thus the combustion needs to be modeled.

3.2.3 Combustion Modeling

The usual models for describing premixed flame propagation include thickened flame technique and flamelet concept such as G-equation/progress variable [24]. In the present work, the premixed combustion is modeled using a thickened flame approach and burning velocity model based on progress variable in the laminar and LES calculations, respectively.

3.2.3.1 Thickened Flame (TF) Model

The TF approach was first proposed by Butler and O'Rourke [25] to capture a laminar premixed flame on a coarser grid. In the TF model the basic idea is that the flame is artificially thickened, to maintain the laminar flame speed S_L, by increasing the diffusivity and decreasing the reaction rate proportionally. According to the theories of laminar premixed flame [26], the laminar flame speed S_L and thickness δ_L can be related to the molecular diffusivity D and mean reaction rate $\bar{\omega}$:

$$S_L \propto \sqrt{D\bar{\omega}}, \ \delta_L \propto \sqrt{D/\bar{\omega}}, \tag{3.16}$$

When the flame thickness is increased by a thickening factor F to include several computational cells, the diffusivity is multiplied by F and the reaction rate is divided by F accordingly without altering the laminar flame speed. The flame is dynamically thickened in a narrowband around the flame front to prevent erroneous mixing away from the flame due to the increased diffusivities. The diffusivity is thus dynamically determined as [27]:

$$D_{\text{eff}} = D(1 + (F - 1)\Omega), \tag{3.17}$$

Ω ranges from unity around the reaction front to zero outside the flame and is given by:

$$\Omega = \tan h\left(\beta \frac{|\bar{\omega}|}{\max(|\bar{\omega}|)}\right), \tag{3.18}$$

where $|\bar{\omega}|$ is the absolute value of reaction rate, β is a constant parameter controlling the thickness of the transition layer between thickened and non-thickened zones. $\max(|\bar{\omega}|)$ is the maximum value of $\bar{\omega}$ in the domain. Therefore, the effective diffusivity is the non-thickened value of laminar diffusivity D away from the flame front. The flame thickness is obtained by [24]:

$$\delta_L = 2\delta(T_2/T_1)^{0.7} = 2\lambda_1/(\rho_1 C_p S_L) \cdot (T_2/T_1)^{0.7}, \tag{3.19}$$

where $\delta = 2\lambda_1/(\rho_1 C_p S_L)$ is the diffusive thickness which is usually 4–5 times larger than a real laminar flame thickness [24]. T_1 and T_2 are the temperature of unburned mixture and the burnt gas, respectively. λ_1, ρ_1, and C_p are the thermal conductivity, density, and specific heat of the unburned mixture, respectively. The premixed flame thickness calculated by Eq. (3.19) is very close to the real flame thickness [24]. The thickening factor is calculated as:

$$F = N\Delta/\delta_L, \tag{3.20}$$

where N is the cell number specified in the flame, Δ is the grid size.

A premixed flame can be resolved by a TF model on a coarse grid. The TF technique can be extended to incorporate with multistep chemistry scheme since the chemical reaction rates are calculated exactly without any ad hoc parameter or submodel [24, 28, 29]. Besides, the phenomena of quenching and ignition can be also simulated by the TF method [24, 29, 30].

3.2.3.2 LES Burning Velocity Model

A three-dimensional CFD code is essential to reproduce the flame behavior and pressure dynamics in a square cross section duct [31], and thus we use an LES turbulent combustion model in order to run simulations within affordable computer resources. Use of chemical kinetics models for combustion simulation would be possible in 2D cases, see e.g. [32] where axisymmetric flame propagation was modeled. However, modeling of experiments on premixed flame propagation even on a relatively small scale remains computationally expensive.

The LES combustion model employed in the present study is the "multi-phenomena combustion model" described in [17]. This combustion model takes into account the following phenomena that influence the burning velocity of the flame: (1) changes of pressure and temperature in the unburned gas; (2) turbulence in the incoming unburned mixture; (3) turbulence generated by flame front itself; (4) preferential diffusion instability or leading point phenomenon. It is based on the assumption of a flamelet regime of premixed turbulent combustion. In the model, the mass burning rate is determined in general as a product of the local burning velocity in a flamelet and the surface area of the flamelet which are both affected by different instabilities.

(1) Effect of pressure and temperature on the burning rate

The values of laminar burning velocity for the whole hydrogen flammable range can be found in [33–36]. The laminar burning velocity reported in [33] is applied in the simulations, e.g., S_{u0} = 2.29 m/s for 35 % of hydrogen by volume in air (T = 298 K, p = 101,305 Pa). The dependence of laminar burning velocity on temperature and pressure during flame propagation is taken into account using an assumption of adiabatic compression [24]:

$$S_L = S_{L0} \cdot \left(\frac{T}{T_{u0}}\right)^{m_0} \cdot \left(\frac{p}{p_0}\right)^{n_0} = S_{L0} \cdot \left(\frac{p}{p_0}\right)^{\varepsilon}, \tag{3.21}$$

where $\varepsilon = m_0 + n_0 - m_0/\gamma$ is the overall thermokinetic index, and γ is the adiabatic index of the unburned mixture. The values of thermal and overall thermokinetic indices for a hydrogen-air mixture with a hydrogen concentration of 35 % by volume are taken from [37] as $m_0 = 1.5$ and $\varepsilon = 0.52$, respectively. The effect of hydrogen concentration on the burning rate is included into the value of the initial laminar burning velocity S_{L0} when it is needed.

(2) Effect of flow turbulence on the burning rate

When turbulence occurs in the flow during combustion wave propagation the flame front would be distorted and the burning rate will be enhanced consequently. The flame thickness is normally a fraction of a millimeter and it is impractical to resolve in simulations the three-dimensional real flame thickness and the flow turbulence at all scales, i.e., it is impossible to perform DNS at comparatively large scales characteristic for safety problems. The effect of flow turbulence on the flame front wrinkling is partially resolved by LES at scales comparable to the mesh applied. The effect of unresolved SGS flow turbulence is modeled following

Yakhot's transcendental equation for the turbulent burning velocity S_t of premixed turbulent combustion [38]:

$$S_t = S_L \cdot \exp(u'/S_t)^2, \qquad (3.22)$$

where u' is the SGS residual velocity. The model is based on the same RNG method [39] and "does not include adjustable parameters" [38]. The paper by Pocheau [40] suggests that Yakhot's equation is not exact but approximate. However, it is also indicated that Yakhot's equation is suitable for a weak turbulence regime, which is typical for accidental deflagrations usually initiated in initially quiescent mixtures similar to the present study. The RNG model was developed for a very thin flame front and was demonstrated to agree well with experimental data in the range $1 \le u'/S_u \le 20 - 25$ [38]. Nevertheless, the model does not take into account phenomena, which are different from the flow turbulence in the incoming flow of unburned mixture, e.g., turbulence generated by flame front itself and various flame instabilities leading to augmentation of the burning velocity. Though their effects are expected to be unimportant in the strong turbulence regime as assumed, for example, in the study [38], for low-turbulence combustion such as at the initial stage of accidental deflagrations they should be taken into account. The transcendental equation for the turbulent burning velocity S_t may be solved iteratively using a successive approximations method; an initial guess $S_t = S_L + 0.5u'$ and an under-relaxation factor of 0.2 provide stable and fast (usually within 5 iterations) convergence of the iterative process. Graphical solution of the RNG turbulent combustion model [38] is given in graphical form in Fig. 3.3.

The Yakhot's original equation [38] is derived for the turbulent burning velocity by the analysis of the effects of both the turbulence in the incoming flow and the laminar burning velocity. It does not take into account in any way other existing physical phenomena, i.e., flame wrinkling due to selective diffusion instability and turbulence generated by the flame front itself, both of which can significantly affect the value of the turbulent burning velocity. These two phenomena need to be

Fig. 3.3 Solution for the RNG turbulent burning velocity in nondimensional form

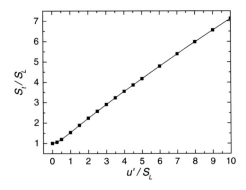

modeled as their contribution to augmentation of mass burning rate can be significant. In order to account for various unresolved physical mechanisms that enhance the burning rate, the laminar burning velocity in the original Yakhot's equation is substituted by a "wrinkled flame burning velocity" S_L^w in the model:

$$S_t = S_L^w \cdot \exp(u'/S_t)^2. \tag{3.23}$$

This is the essential modification of the original Yakhot's equation. It is worth noting that S_L^w does not include the effect of the turbulence in the incoming unburned mixture. The wrinkled flame burning velocity S_t^w influences the total turbulent burning rate through the interaction with the flow turbulence in the unburned mixture that is calculated by the modified Yakhot's Eq. (3.23).

(3) Effect of turbulence generated by flame front itself

The turbulent flame front generates additional turbulence in the near-field region [41]. The increase of surface area of the flame front due to the turbulence generated by the flame front itself cannot be resolved with current computing power even for moderate scale practical problems. The upper limit for a flame wrinkling factor due to the self-induced turbulence can be derived for high Reynolds number flows as [42]:

$$\Xi_K^{\max} = (E - 1)/\sqrt{3}, \tag{3.24}$$

where E is the expansion coefficient or ratio of densities of the unburned mixture and the combustion products. It is $E = 6.82$ for a 35 % hydrogen-air mixture. For a flame propagating in an initially quiescent mixture, this wrinkling factor gradually increases from a value of one at the ignition point to the maximum value of $\Xi_K^{\max}$ for fully developed turbulence. Gostintsev et al. [43] observed that the critical radius for the onset of self-similar flame propagation (fully developed turbulence) for near-stoichiometric hydrogen-air mixtures is about $R_0 = 1.0 - 1.2$ m. In order to take into account these transitional effects, the following equation is applied for SGS modeling of the unresolved self-induced turbulence:

$$\Xi_K = 1 + \left(\psi \cdot \Xi_K^{\max} - 1\right) \cdot [1 - \exp(-R/R_0)], \tag{3.25}$$

where R is the distance from the ignition point to the flame front, and $\psi \leq 1$ is an "ad hoc" model constant reflecting the extent to which the theoretical maximum value of the turbulence generated by flame front itself is reached. In previous studies the value of the constant for near-stoichiometric mixtures is close to $\psi = 0.5$ and it grows to maximum value $\psi = 1$ for lean hydrogen-air mixtures [44]. In the present case of 35 % hydrogen-air mixture the accepted constant is $\psi = 0.6$. The experimental value of the critical radius $R_0 = 1.2$ m is adopted in the simulations. Therefore, with consideration of the physical phenomenon of the turbulence generated by flame front itself the following expansion for the wrinkled flame burning velocity is obtained:

$$S_L^w = S_L \cdot \Xi_K. \tag{3.26}$$

(4) Effect of preferential diffusion coupled with flamelet curvature

A hydrogen flame with a Lewis number less than unity is affected by the preferential diffusion. For a given mixture composition, there exists a flame curvature radius which corresponds to the maximum mass burning rate. Following Zeldovich's idea flamelets of such curvature will lead to turbulent premixed flame propagation. Zimont and Lipatnikov [36], based on the work of Kuznetsov et al. [45], calculated a leading point coefficient Ξ_{lp} associated with this mechanism as a correction to the laminar burning velocity dependent on mixture composition. In the present model, it is assumed that the diffusive-thermal instability develops linearly to the maximum value Ξ_{lp} at half of the critical radius R_0:

$$\Xi_{lp} = 1 + \frac{\left(\Xi_{lp}^{max} - 1\right) \cdot 2R}{R_0}, \tag{3.27}$$

where $\Xi_{lp}^{max} = 1.1$ is the maximum leading point coefficient for a 35 % hydrogen-air mixture. For $R > R_0$, $\Xi_{lp} = \Xi_{lp}^{max}$. In the present simulations, the leading point coefficient never reaches its maximum value because the length scale of the duct is smaller than half of R_0. Taking into account the preferential diffusion mechanism requires the equation for the wrinkled flame burning velocity to be changed to:

$$S_L^w = S_L \cdot \Xi_K \cdot \Xi_{lp}. \tag{3.28}$$

Note that this flame acceleration mechanism plays a relatively small role in a rich hydrogen-air flame, but we retain it for the sake of keeping the general combustion model intact even in rich hydrogen-air flame simulations.

(5) The equation for turbulent burning velocity

As a result of the SGS modeling described above, the multi-phenomena combustion model in this study includes four different physical mechanisms affecting the turbulent burning velocity, i.e., dependence of laminar burning velocity on transient pressure and temperature (incorporated into the burning velocity S_L), turbulence generated by flame front itself and the leading point mechanisms (incorporated in the wrinkled flame burning velocity S_L^w), and flow turbulence in the incoming unburned mixture (simulated in Yakhot's turbulent combustion model). The final expression for the turbulent burning velocity is:

$$S_t = S_L \cdot \Xi_K \cdot \Xi_{lp} \cdot \exp\left(\frac{u'}{S_t}\right)^2 = S_L^w \cdot \exp\left(\frac{u'}{S_t}\right)^2. \tag{3.29}$$

3.3 Numerical Results and Discussion

3.3.1 Results Based on Thickened Flame Technique and Comparisons to Experiments

3.3.1.1 Numerical Methods, Initial and Boundary Conditions

(1) 2D numerical simulations

In the CFD simulations, the TF model described in Sect. 3.2.3.1 is used. There are numerous chemical reaction mechanisms of hydrogen-air reaction [46–48]. The detailed chemistry developed by Miller et al. [48] is adopted. It is composed of 9 species (H_2, O_2, H_2O, OH, O, H, HO_2, H_2O_2, N_2) and 19 elemental reactions, as shown in Table 3.1. Figure 3.4 shows the structure of a one-dimensional steady-state stoichiometric hydrogen-air flame obtained using the detailed chemistry scheme at initial temperature $T_0 = 298.15$ K and pressure $p_0 = 101325$ P (Huahua Xiao and Arief Dahoe, 2011). The calculated adiabatic flame temperature,

Table 3.1 19-step detailed chemical mechanism of hydrogen-air mixture by Miller et al. [48]. Units are moles, cubic centimeters, seconds, Kelvins, and calories

No.	Chemical reaction	Pre-exponential factor	Temperature index	Activation energy
1	$H_2 + O_2 = OH + OH$	1.70E + 13	0.00	47780.0
2	$OH + H_2 = H + H_2O$	1.17E + 09	1.30	3626.0
3	$H + O_2 = O + OH$	5.13E + 16	−0.816	16507.0
4	$O + H_2 = H + OH$	1.80E + 10	1.00	8826.0
5	$H + O_2 + m = HO_2 + m^a$	2.10E + 18	−1.00	0.0
6	$H + O_2 + O_2 = HO_2 + O_2$	6.70E + 19	−1.42	0.0
7	$H + O_2 + N_2 = HO_2 + N_2$	6.70E + 19	−1.42	0.0
8	$HO_2 + OH = H_2O + O_2$	5.00E + 13	0.00	1000.0
9	$HO_2 + H = OH + OH$	2.50E + 14	0.00	1900.0
10	$HO_2 + O = OH + O_2$	4.80E + 13	0.00	1000.0
11	$OH + OH = H_2O + O$	6.00E + 08	1.30	0.0
12	$H_2 + m = H + H + m^b$	2.23E + 12	0.50	92600.0
13	$O_2 + m = O + O + m$	1.85E + 11	0.50	95560.0
14	$H + OH + m = H_2O + m^c$	7.50E + 23	−2.60	0.0
15	$HO_2 + H = H_2 + O_2$	2.50E + 13	0.00	700.0
16	$HO_2 + HO_2 = H_2O_2 + O_2$	2.00E + 12	0.00	0.0
17	$H_2O_2 + m = OH + OH + m$	1.30E + 17	0.00	45500.0
18	$H_2O_2 + H = H_2 + HO_2$	1.60E + 12	0.00	3800.0
19	$H_2O_2 + OH = H_2O + HO_2$	1.00E + 13	0.00	1800.0

[a]Third body efficiencies: k_5 (H_2O) = 21 k_5 (Ar), k_5 (H_2) = 3:3 k_5 (Ar)
[b]Third body efficiencies: k_{12} (H_2O) = 6 k_{12} (Ar), k_{12}(H) = 2 k_{12} (Ar), k_{12} (H_2) = 3 k_{12} (Ar)
[c]Third body efficiency: k_{14} (H_2O) = 20 k_{14} (Ar)

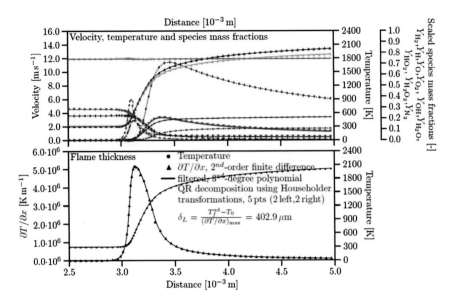

Fig. 3.4 Structure of one-dimensional stoichiometric hydrogen-air flame

laminar burning velocity, and laminar flame thickness are $T = 2386$ K, $S_L = 2.1$ m/s, and $\delta_L = 0.4$ mm, respectively. In the 2D computations, the flame thickness in the TF model is assumed to be a constant.

The governing equations are solved using a finite volume method. A uniform structured grid is adopted with the grid size 0.4×0.4 mm, and 273,156 cells are generated in the entire computation domain. The thickening factor is $F = N \times \Delta/\delta_L$, where N and Δ are the cell number and cell size, respectively. The time integration is advanced by an implicit scheme. The time step Δt is chosen as half of the value derived from the Courant–Friedrichs–Lewy criterion: $\Delta t = (\Delta/c_s)/2 = 5.0 \times 10^{-7}$ s to capture the pressure wave in the duct, where $c_s = 400$ m/s is the sound speed in a 30 % hydrogen-air mixture. The time step determined by this method is sufficiently small to investigate the coupling of flame with the acoustic waves [49]. The adiabatic and nonslip boundary conditions are applied on the sidewalls. The initial conditions are identical to those in the experiment. A 30 % hydrogen-air mixture is used in the 2D numerical simulations.

(2) 3D numerical simulations

Similar to the 2D calculation, the governing equations are discretized by using a finite volume method. The third-order MUSCL scheme was used for convection terms. The base grid for the numerical simulation is a 3D uniform structured grid with cell size $4 \times 4 \times 4$ mm, and covers the entire domain. Three additional levels of dynamically and locally adaptive mesh refinement are adopted, and track the location of the flame front. Each successive level is a factor of 2 finer, yielding a grid resolution of 0.5 mm around the reaction zone. The TF technique thickens the flame front so that it is resolved on the computational grid [29]. Approximately 5–8 grid points are generally required in order to resolve the thickened structure of the premixed flame [29]. The flame thickness and thickening factor are calculated using Eqs. (3.19) and (3.20).

In order to save computational cost, the chemical reactions of hydrogen in air are taken into account using a simplified seven-step chemistry scheme with six species (H_2, O_2, H_2O, OH, O, H) by Lacaze et al. [29] instead of the detailed chemistry described above. The scheme is given in Table 3.2. According to Lacaze et al. [29], his scheme is able to accurately reproduce the laminar burning velocity and adiabatic flame temperature over a large range of equivalence ratios (especially for lean and stoichiometric conditions) and to account for pressure effects. The same equivalence ratio of hydrogen-air mixture is used as that in 2D. The obtained initial laminar flame thickness and laminar burning velocity are δ_L = 0.41 mm and S_{L0} = 2.05 m/s, respectively. The gravity effect can be neglected in the numerical simulation with Froude number $Fr = S_L^2/(gH) > 10$ in the flame propagation, where H = 4.1 cm is half of the height of the duct and g = 9.8 m/s^2 is the acceleration of gravity.

The mixture is quiescent before ignition. The initial temperature and pressure are the same as those in the experiment. The no-slip adiabatic boundary condition is used on the sidewalls. The ignition location is identical to that in the experiment. The geometry of the tube matches the experimental combustion duct.

3.3.1.2 Results and Discussion of 2D Numerical Simulations

Here the experimental results are first presented to provide data for comparisons between numerical simulation and experiment. Figure 3.5 shows a sequence of high-speed schlieren images of premixed hydrogen-air flame propagating in the closed duct, illustrating the development of tulip and distorted tulip flames. For the distorted tulip flame, five stages of the flame dynamics can be distinguished based on the four stages of classical tulip flame propagation proposed by Clanet and Searby [50]: spherical flame, finger-shape flame, flame with the skirt touching the

Table 3.2 Seven-step hydrogen-air chemical reaction kinetics by Lacaze et al. [29]. Units are moles, cubic centimeters, seconds, Kelvins, and calories

No.	Chemical reaction	Pre-exponential factor	Temperature index	Activation energy
1	$H + O_2 = O + OH$	3.62E + 17	−0.91	1.653E + 04
2	$O + H_2 = H + OH$	1.53E + 05	2.67	6.296E + 03
3	$H_2 + O_2 = OH + OH$	5.13E + 13	0.00	4.805E + 04
4	$OH + H_2 = H_2O + H$	6.64E + 13	0.00	5.155E + 03
5	$OH + OH = H_2O + O$	1.90E + 13	0.00	1.091E + 03
6	$H + OH + m = H_2O + m$	6.67E + 22	−2.00	0.000
7	$H + H + m = H_2 + m$	2.20E + 18	−1.00	0.000

Third body efficiencies: 2.5 for H_2, 16 for H_2O, and 1.0 for all others

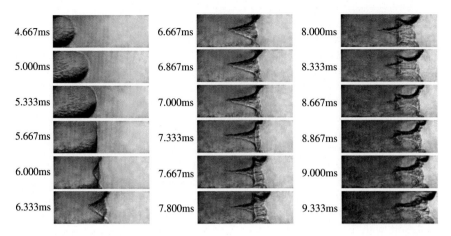

Fig. 3.5 A sequence of frames from a high-speed schlieren movie of premixed hydrogen-air flame in a closed duct

sidewalls, tulip flame, and distorted tulip flame. In this study, main attention is given to the later stages of the flame dynamics, especially on the new stage of distorted tulip flame. The flame front is flattened at $t = 5.667$ ms. Thereafter, a well-pronounced classical tulip flame is produced. The distortions are initiated in the immediate proximity of the flame leading tips at around $t = 7.0$ ms. Subsequently, two secondary cusps are generated on the two primary lips, forming a noticeable distorted tulip flame, e.g., the flame at 7.667 ms. When the distortions locate near the center of the primary tulip lips the distorted tulip flame develops into a "triple tulip" shape with the two secondary cusps comparable to the primary one (e.g., the flame shape at $t = 7.8$ ms). The "triple tulip" flame lasts for a short time.

During the formation of the classical tulip flame, the primary cusp moves backward while the flame leading tip propagates forward, creating a deep tulip cusp. However, the primary tulip cusp travels forward after the generation of the distortions. On the contrary, the distortions move backward at a relatively small speed. Finally, the primary distorted tulip flame disappears with the primary cusp and distortions propagating one to the other, e.g., at $t = 9.333$ ms. Just before the collapse of the primary distorted tulip flame, two additional distortions are created near the tips of the primary tulip lips after $t = 8.333$ ms, resulting in a second distorted tulip flame. The second distorted tulip flame develops with a cascade of distortions superimposed on the primary lips (e.g., $t = 8.867$ ms). As the primary distortions vanish the secondary distortions almost arrive at the center of the primary lips. Actually, the distorted tulip repeats itself twice with a series of distortions created on the primary lips.

The development of the premixed flame in the numerical simulation is presented in Fig. 3.6, showing the typical flame shapes at the aforementioned five stages. The numerical calculation successfully reproduces the five stages of the flame dynamics. Similar to the experimental observation, the flame expands spherically after ignition.

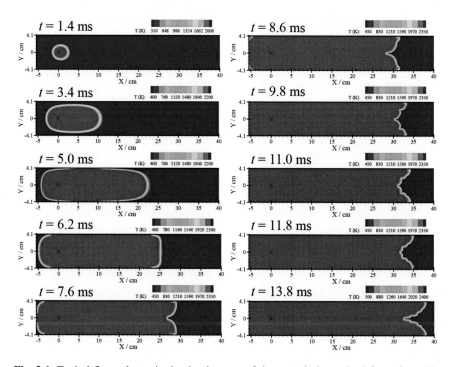

Fig. 3.6 Typical flame shapes in the development of the numerical premixed flame front. The *colors* represent the temperature from the unburned mixture (*blue*) to the burnt gas (*red*)

The finger-shaped flame starts at around $t = 1.4$ ms. The flame front touches the sidewalls at about $t = 4.8$ ms. The contact of the flame with the sidewalls leads to an abrupt flame deceleration, during which a planar flame front is created (see the flame at $t = 6.2$ ms). A classical tulip flame is formed subsequent to the flame front inversion. The distorted tulip flame is initiated at approximately $t = 7.6$ ms immediately after another sudden deceleration of the flame front. The distortions originate from the tips of the primary tulip lips and move toward the primary tulip cusp. A well-pronounced distorted tulip flame ("triple tulip" flame) is formed as the distortions locate near the center of the primary tulip lips. The well-pronounced distorted tulip flame appears very close to that in the experiment with two noticeable secondary cusps on the primary tulip lips (e.g., the flame shape at $t = 9.8$ ms). The first distorted tulip flame ultimately vanishes with the distortions and the primary tulip cusp propagating one toward the other. Additional smaller distortions can also be seen on the flame front after the full formation of the distorted tulip flame (e.g., the flame at $t = 11.8$ ms). The second distorted tulip flame in the numerical simulation is not as noticeable as that in the experiment. This discrepancy is due to the artificially thickened numerical flame front. Excessively thickened flame front makes the distorted tulip flame less pronounced. Actually, the second distorted tulip flame is much less pronounced than the first one even in the experiment, as shown in Fig. 3.5.

Overall, it can be deduced from the above analysis that the numerical simulation based on a TF model reasonably reproduces the combustion dynamics of the same experiment. Particularly, the flame front evolution in the numerical simulation is in good agreement with those observed in the experiment, including the initiation, development, and collapse of the distorted tulip flames.

3.3.1.3   Results and Discussion of 3D Numerical Simulations

(1) The development of flame front
For the flame propagation in a duct with a square cross section, 3D numerical simulation may be more preferable than 2D [31]. Nevertheless, 3D calculation is also much more expensive (sometimes even impractical) if the same grid resolution as in 2D is used.

Predicted flame front shapes in the closed duct at different time instants in the 3D numerical simulation using the TF technique are given in Fig. 3.7. The four stages of the flame dynamics observed in the experiment are reasonably reproduced in the numerical simulation. The spherical flame (see Fig. 3.7a) lasts approximately 1.2 ms in the numerical simulation which is equal to that in the experiment. At the first stage, the flame expands freely in a short time and the burning velocity is very close to the initial laminar burning velocity. This stage can be predicted very well in the numerical simulation. The flame develops a finger-shaped front after spherical stage, as shown in Fig. 3.7b. The flame skirt starts to touch the sidewall at about $t = 4.8$ ms in the numerical simulation which is slightly larger than that in the experiment. In fact, at the second stage the flame front start to be wrinkled and more

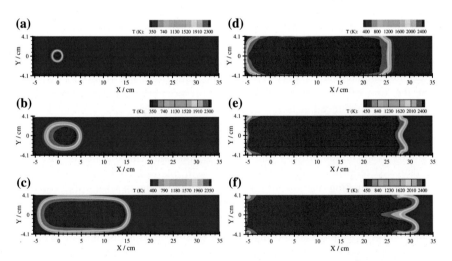

Fig. 3.7 Numerical flame shape at the central plane of the duct at times $t = 0.8, 2.8, 4.8, 6.2, 7.4,$ and 8.6 ms. The *colors* designate the temperature from the unburned mixture (*blue*) to the burnt gas (*red*)

wrinkles are generated as the flame continues to propagate. The laminar burning velocity is consequently enhanced by the effect of wrinkling. However, the real flame is wrinkled at scales below the resolution in the numerical simulation. The laminar burning velocity is thus underestimated in the numerical simulation. In the numerical simulation the flame front is flattened at about $t = 6.2$ ms which is also slightly larger than that in the experiment, as expected. With respect to the evolution of the flame shape, the numerical simulation reproduces the experiment well, especially for the first four stages.

(2) The dynamics of flame leading tip

Figure 3.8 presents the experimental and numerical location of the flame leading tip as a function of time. The position of the flame leading tip is defined as the distance from the ignition point. Before flame inversion the flame tip at the centerline of the duct is taken as the flame leading tip, while after inversion the tip of the upper part of the flame was treated as the flame leading tip both in the experiment and numerical simulation. Figure 3.9 shows the experimental and predicted propagation speed of the flame leading tip. At the spherical flame stage, the flame expands at a constant speed of $E \cdot S_L$. The flame accelerates fast at the second stage with its surface area increasing exponentially. The flame acceleration stops when the flame skirt reaches the sidewalls. Meanwhile, the flame surface area

Fig. 3.8 Experimental and 3D numerical location of the flame leading tip

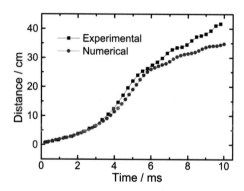

Fig. 3.9 Experimental and numerical propagation speed of the flame leading tip

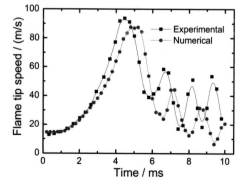

decreases rapidly because a part of the flame front near the sidewalls vanishes. The expansion of the combustion products also decreases since fewer burnt gases are produced. Therefore, the flame front decelerates quickly, as shown in Fig. 3.9.

Oscillations develop in the trajectories of both the position (Fig. 3.8) and propagation speed (Fig. 3.9) of the flame leading tip after the flame skirt reaches the sidewalls. Both the period and the amplitude of the oscillations in the numerical simulation are in close agreement with that in the experiment. In the initial stages, i.e., the spherical stage and the early period of the finger-shape stage, the flame propagates in the numerical simulation almost at the same speed as in the experiment. After $t = 2.4$ ms, the predicted flame moves more slowly than that in the experiment. The discrepancy of both the location and speed of the flame leading tip between the experiment and the numerical simulation becomes more noticeable after the flame inversion (see Figs. 3.8 and 3.9). As pointed out above, this behavior is mainly caused by the fact that the effect of the flame wrinkles is not resolved by the TF model on the grid resolution in the present numerical simulation. The laminar burning velocity is consequently underestimated in the numerical simulation as the wrinkle grows at the finger-shape stage. In the initial stages, the difference between the numerical and experimental results is small, which suggests that the wrinkle of the flame front play a relatively minor role in the flame dynamics in these stages. Nevertheless, the contribution of the flame wrinkling to the burning velocity augmentation can be more considerable in the later stages of the flame propagation [21, 51]. A lower pressure in the unburned mixture is achieved in the later stages because the pressure buildup is related to the laminar burning velocity. The pressure rise in the unburned gas can considerably enhance the laminar burning velocity in turn through the compression (see Eqs. (1.6) and (1.8)). Therefore, the laminar burning velocity will be further underestimated in the numerical simulation as the pressure difference grows larger. Overall, the agreement between the predicted and experimental flame dynamics is good, especially before the flame inversion.

(3) The pressure dynamics inside the duct

Figure 3.10 presents the experimental pressure dynamics obtained from the pressure transducer and the 3D simulated pressure dynamics recorded at the same location. In the experiment the sudden drop of pressure growth at about $t = 12.1$ ms corresponds to the opening of the discharge vent. At the first stage the pressure in the duct does not increase, as shown in Fig. 3.10, and has little influence to the laminar burning velocity. The pressure begins to increase fast with the flame surface area increasing exponentially at the finger-shape stage both in the experiment and numerical simulation. After the flame skirt touches the sidewalls the pressure growth rate drops due to the sudden reduction of the generation of combustion products which results from the drastic reduction of the flame surface area. This result is similar to that in [50, 52]. In addition, a pressure wave is triggered by the first contact of the flame skirt with the lateral walls of the duct. The flame surface

Fig. 3.10 Comparison
between experimental and 3D
simulated pressure dynamics

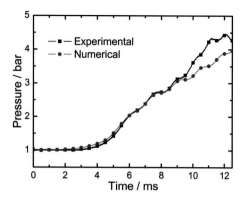

area reaches its minimum value when the flame front is flattened at $t = 6.067$ ms in
the experiment and $t = 6.2$ ms in the numerical simulation, respectively. The growth
rate of the pressure begins to increase after the flame inversion due to the development of the tulip flame which consequently leads to an increase in the flame
surface area.

Obvious pressure oscillations occur both in the experiment and numerical simulations after the flame reaches the sidewalls of the duct. The predicted period of the
pressure oscillations is consistent with that in the experiment. Both the appearance
of the plateau period and the oscillation period of the pressure trajectory (Fig. 3.10)
coincide with the sudden deceleration of the flame leading tip (Fig. 3.9). According
to Gonzalez [49], the oscillations of both the flame front and the pressure rise have a
close connection with the pressure wave (initiated as the flame touches the lateral
walls). Therefore, it can be concluded that the pressure wave plays an important
role in the dynamics of both the flame front and the pressure during the premixed
hydrogen-air flame propagation in the closed duct. The pressure in the numerical
simulation is pretty close to that in the experiment in the early stages. This confirms
further that the wrinkle of the flame front has a relatively smaller effect on the flame
acceleration in the early stages. After the flame inversion the pressure in the
numerical simulation increases more slowly than in the experiment as the tulip
flame grows. This circumstance is also due to the effects of the wrinkle is not taken
into account in the numerical simulation. The agreement between the numerical
simulation and the experiment is good, especially for the early stages, although
there are some discrepancies after the appearance of the tulip flame.

Note that although no noticeable distorted tulip flame phenomenon is observed
in the 3D numerical simulation, the major features of flame tip and pressure
dynamics are consistent with those of a distorted tulip flame. This implies that the
inherent characteristics of premixed hydrogen-air flame propagation in the duct
have been reproduced well. It also indicates that the periodic oscillating flame
behavior is independent of the formation of distorted tulip flame.

3.3.2 LES Calculations Using Burning Velocity Model and Comparisons to Experiments

3.3.2.1 Numerical Methods, Initial and Boundary Conditions

The LES approach described in Sect. 3.2.2.2 together with the burning velocity model in Sect. 3.2.3.2 are used here to simulate hydrogen-air flame propagation in a closed duct. Again, the equations are discretized using a finite volume method. Two different grids are used in order to examine the grid effect of the 3D simulations. Grid 1 is a 3D nonuniform structured grid with refinement near the boundaries. The size of the cells closest to the walls is 0.1 mm to keep dimensionless wall distance $y^+ < 10$ at the first control volume, where $y^+ = \Delta y \cdot \rho \cdot u_\tau / \mu$, Δy is the distance to the nearest wall, u_τ is the friction velocity at the nearest wall. The characteristic grid size in the main domain of the flow field is 2 mm and 1,085,440 hexahedral cells are generated in the computational domain. Grid 2 is a 3D uniform structured grid composed of 58,653 hexahedral cells with a characteristic grid size of 4 mm throughout the entire domain.

The initial temperature and pressure in the simulations are kept the same as in the experiment. The mixture, which is initially quiescent, is a 35 % hydrogen-air mixture. The laminar burning velocity is $S_{L0} = 2.29$ m/s at $T_0 = 298$ K and $p_0 = 101,305$ Pa [33]. The expansion ratio is $E = 6.82$. The temperature and thermokinetic indices are $m_0 = 1.5$ and $\varepsilon = 0.52$ [37], respectively. The model constant and critical radius are $\psi = 0.6$ and $R_0 = 1.2$, respectively. The initial progress variable is equal to zero all over the computational domain. The ignition location matches that in the experiment. Combustion is initiated by increasing the progress variable from zero to one in a small ignition domain during a period of 0.1 ms. This ignition duration is calculated by dividing the radius of the ignition domain by the propagation velocity $S_L \cdot E$. The geometry of the duct in the numerical simulations is identical to that in the experiment. For convenience of analysis in the simulations the origin of the coordinate system is located at the ignition point, i.e., in simulations the ignition point is at location $x = 0$ cm and the left-hand end of the duct is at $x = -5.5$ cm. A no-slip adiabatic boundary condition is used on all the walls.

Simulations are performed on the CFD platform FLUENT 6.3.26. The parallel solver of double precision is employed with explicit linearization of the LES governing equations. The third-order MUSCL scheme is applied for convection terms. The Courant–Friedrichs–Lewy (*CFL*) number is equal to *CFL* = 0.8 to ensure stability. A simulation of 10 ms of real time takes 18 days on a 8 core Fujitsu workstation PRIMERGY TX300 S5 (grid 1).

3.3.2.2 Results and Discussion of LES Numerical Simulations

(1) Flame front development and shape changes in the closed duct

The experimental results of 35 % hydrogen-air flame propagating in the closed duct are presented first here for comparisons to LES simulations. A sequence of high-speed schlieren images during hydrogen-air flame propagation in the tube is presented in Fig. 3.11. Figure 3.11a shows the early stages of the flame propagation, while Fig. 3.11b displays the later stages. The experiment is repeated five times for both the early and later stages. The results indicate that the flame shape, pressure transients, and flame arrival times are quite reproducible.

It has been known in Sect. 3.3.1.2 that the flame propagation with distorted tulip shape can be divided into five stages. Together with the "distorted tulip" flame stage the present experimental flame dynamics in this study are as follows: (1) spherical flame, $0 < t < 1.067$ ms, this first stage begins after ignition and the flame expands spherically, unaffected by the sidewalls; (2) finger-shaped flame, 1.067 ms $< t < 3.667$ ms, the flame front approaches the sidewalls and changes from spherical into finger shaped with an exponential increase of the flame surface; (3) flame with the skirt touching the sidewalls, 3.667 ms $< t < 4.867$ ms, the lateral sides of the elongated flame quenches gradually near the walls and the flame surface

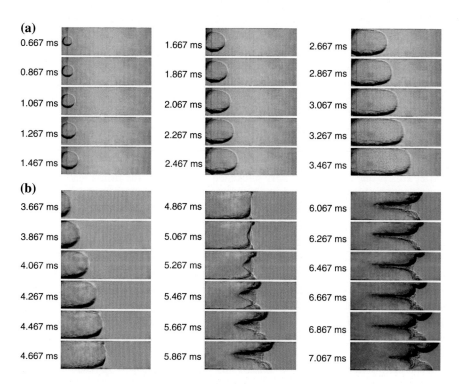

Fig. 3.11 High-speed schlieren images of premixed hydrogen-air flame propagation: **a** early stages, **b** later stages

area begins to decrease; (4) tulip flame, 4.867 ms < t < 6.0 ms, occurs after the inversion of the flame front at t = 4.867 ms, during this stage, the flame front develops with a cusp pointing backward to the burnt gas and a tulip shape is formed consequently; (5) "distorted tulip" flame, t > 6.0 ms, at this new fifth stage, the tulip flame becomes distorted with the original tulip lips dented and concaved toward the sidewalls by secondary cusps (indentations).

Starting from the second stage the experimental flame front begins to be obviously wrinkled and flame instabilities generate more wrinkles as the flame propagates. This increases the flame surface area. After the flame front is flattened at t = 4.867 ms, the original tulip cusp keeps propagating backward (in a system of coordinates fixed to the duct) and a deep cusp is formed, e.g., the flame at t = 5.867 ms. Meanwhile, the leading flame front propagates forward. The different propagation directions between the cusp and the leading flame front produce a very pronounced tulip shape at the fourth stage. The original tulip cusp propagates forward after the "distorted tulip" flame is well established. This is due to the flow pattern formed in the duct as we will see below in sections regarding to flame–flow interactions.

In the experiment, the distortions are initiated at around t = 6.0 ms very close to the tip of the original tulip lips. As the "distorted tulip" flame develops, a smaller secondary cusp grows gradually on each of the lips after the tulip flame is fully developed (see Fig. 3.11b). The secondary cusps move backward at a relatively low speed and become more and more pronounced as the leading front of the flame propagates forward at time 6.0 ms < t < 6.867 ms, as shown in Fig. 3.11b. After t = 6.867 ms, the secondary cusps start to travel forward together with the flame leading front. Ultimately, the "distorted tulip" flame gradually develops into a "triple tulip" flame as the secondary cusps move to the center of the primary lips, as shown in Fig. 3.11b.

The numerical simulations are performed on two grids. The numerical tulip flame simulated on courser grid 2 is less pronounced than that on finer grid 1, as expected. No significant "distorted tulip" flame can be reproduced on the courser grid 2. This is due to an excessively thickened numerical flame front created by the courser grid, which causes the distortion to be "smoothed out." The finer grid reproduces a deeper cusp and more notable tulip shape. Figure 3.12 demonstrates shapes of the premixed hydrogen-air flame at the central plane of the combustion duct (grid 1). The typical stages of the flame shape evolution are illustrated at time instants t = 0.9218 ms (a), 3.6402 ms (b), 4.4175 ms (c), 5.2977 ms (d), 6.5267 ms (e), 7.7371 ms (f). The five stages observed in the experimental flame propagation are reproduced in the simulations on grid 1. The tulip flame and "distorted tulip" flame are initiated at about t = 5.2977 ms and t = 6.8747 ms, respectively, slightly later than in the experiment. The numerical "distorted tulip" flame is shown in Fig. 3.12f, it is less pronounced than that in the experiment. Figure 3.13 shows the 3D "distorted tulip" flame at t = 7.7371 ms in comparison with a well-pronounced classical tulip flame shape at t = 6.5267 ms (iso-surface c = 0.5). The "distorted tulip" flame is more distinguished in the 3D view (Fig. 3.13a, b) than in the 2D view in the central plane cross section (Fig. 3.12f).

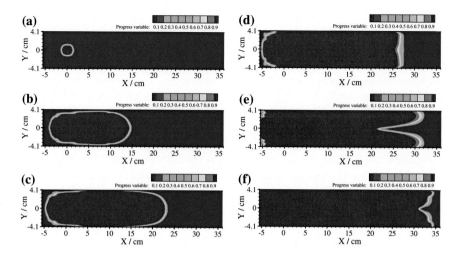

Fig. 3.12 Numerical flame shape at the central plane of the combustion duct at $t = 0.9218$ ms (**a**), 3.6402 ms (**b**), 4.4175 ms (**c**), 5.2977 ms (**d**), 6.5267 ms (**e**), 7.7371 ms (**f**). The *colors* designate the progress variable from the unburned gas (*blue*) to the combustion products (*red*)

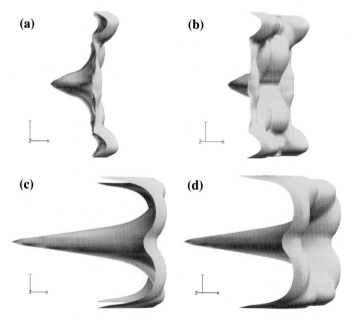

Fig. 3.13 "Distorted tulip" flame at $t = 7.7371$ ms: **a** lateral view, **b** oblique view. Classical tulip flame at $t = 6.5267$ ms: **c** lateral view, **d** oblique view

(2) The dynamics of flame front

Figure 3.14 shows the position and propagation speed of the flame leading tip in a comparison between experiment and LES simulations, where location is defined as the distance from the ignition point. The quantities Z_{tip} and S_{tip} are the physical positions and propagation speed of flame leading tip. For comparison with the theory below the length scale, timescale and flame propagation speed are reduced by H, H/S_{L0} and S_{L0} respectively, where half of the width of the duct is $H = 4.1$ cm and the initial burning velocity $S_{L0} = 2.29$ m/s as used in LES simulations. Figure 3.15 presents a comparison of the experimental and numerical position and propagation speed of the primary tulip cusp. In the figure Z_{cusp} and S_{cusp} denote the physical location and propagation speed of the primary tulip cusp.

Oscillations are seen in the trajectories of both the locations (Fig. 3.14a) and speed (Fig. 3.14b) of the flame leading tip, and the flame is periodically "stagnant" both in the experiment and the numerical simulations after the flame touches the sidewall. The flame propagates slightly faster in the experiment than in the

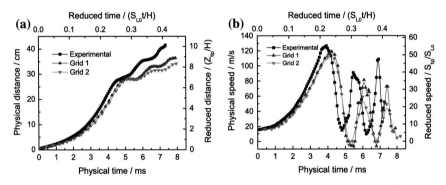

Fig. 3.14 Location (**a**) and propagation speed (**b**) of the flame leading tip in the experiment and LES simulation

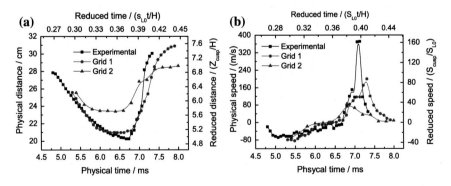

Fig. 3.15 Experimental and numerical location (**a**) and propagation speed (**b**) of the primary tulip cusp

simulations before flame inversion. After the inversion the discrepancy between experimental and numerical results becomes more noticeable. This is thought to be due to the fact that in the experiment there is a more pronounced tulip flame shape with a larger flame surface area and thus a higher burning rate is achieved. The flame simulated on grid 1 propagates almost at the same speed as that simulated on grid 2 in the first three stages and then slightly faster. The reason for this is that, as pointed out previously (Sect. 3.2.3.2), the mass burning rate simulated by the gradient method is independent of grid resolution (unless the number of cells in the numerical flame front decreases when it touches a wall). This ensures that before the flame inversion the convex numerical flame propagates nearly at the same speed on both grids.

However, after the flame inversion the simulation on grid 1 creates a deeper tulip cusp, as shown in Fig. 3.15a, a more pronounced tulip shape and a "distorted tulip" flame with a larger flame surface. The predicted cusp position using grid 1 is reasonably close to that in the experiment while the simulation result based on grid 2 is poorer because of the coarser grid resolution. Figure 3.15b shows that the primary tulip cusp propagates backward first (until about 6.73 ms in the experiment) and then accelerates forward at a higher speed (Fig. 3.15b). In the experiment the speed of the primary cusp reaches its maximum value around $t = 7.067$ ms, at which time the "distorted tulip" flame is well established (Fig. 3.11b). The sudden increase in the speed of the primary cusp is due to the violent consumption of the fuel mixture confined in the elongated cusp (with propagation of the primary tulip lips one toward the other). This phenomenon is reasonably captured by the simulation on grid 1, though the acceleration of the cusp is smaller than that in the experiment due to the relatively thicker numerical flame front. The difference between the numerical result on grid 2 and experiment is relatively large, as shown in Fig. 3.15b.

(3) The pressure dynamics

In the above experiments, the onset of significant flame deformations, i.e., tulip flame and "distorted tulip" flame, coincides with the sudden deceleration of the flame leading tip. Figure 3.16 shows the experimental pressure dynamics obtained from a transducer located at distance $x = 40$ cm and simulated pressure dynamics recorded at the same place. In the experimental results a sudden drop of pressure at about $t = 9.6$ ms corresponds to opening of the discharge vent.

Before the flame touches the sidewalls, the flame leading tip accelerates quickly and the flame surface area increases exponentially both in the experiment and numerical simulations. During this period the internal pressure increases exponentially and monotonically as well. After the flame skirt touches the sidewalls, the flame leading tip decelerates and the pressure growth rate drops due to the reduction of the flame surface area which subsequently results in the reduction of the generation of combustion products. This initiates a pressure wave. The flame surface area reaches its minimum when the plane flame front appears at $t = 4.867$ ms in the experiment, at $t = 5.2977$ ms for simulations on grid 1 and at $t = 5.405$ ms for simulations on grid 2. After the flame inversion, the combustion rate accelerates

Fig. 3.16 Experimental and
LES pressure dynamics

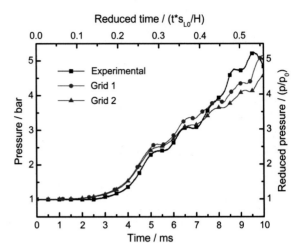

again due to the developing tulip shape and the consequent increase of the flame surface area. Just before the initiation of the "distorted tulip" flame at $t = 6.0$ ms in the experiment and $t = 6.8747$ ms in the simulation (grid 1), the flame leading tip decelerates abruptly once again and the rate of the pressure rise also drops.

Although no apparent "distorted tulip" flame is reproduced in simulations on grid 2, the deceleration of the flame leading tip and the slowdown in the pressure rise caused by the traveling acoustic wave are also observed. This implies that the flame deceleration is an inherent phenomenon of the flame propagation coupled with the acoustic pressure wave (initiated by the first deceleration when the flame touches the wall) traveling back and forth in the duct. The primary tulip cusp accelerates backward to the deepest position and then accelerates forward, as shown in Fig. 3.15.

Pressure oscillations occur both in the experiment and numerical simulations after the flame inversion and their period in simulations is consistent with that in the experiment. According to Leyer and Manson [53], the observed pressure oscillations are directly caused by the sudden reduction of the amount of expanding burnt gas when the flame touches the wall. Before $t = 7.9$ ms and $t = 7.3$ ms in simulations on grid 1 and 2, respectively, the numerical pressure transient is slightly higher than the experimental pressure. The explanation for this deviation is heat losses due to the cold walls in the experiment, which are not modeled. The experimental pressure is higher than simulated as there is a more pronounced "distorted tulip" flame created in the experiment compared to the simulations.

(4) Influence of the wrinkling factor on the combustion dynamics

The LES combustion model used here, i.e., the multi-phenomena combustion model, accounts for four different physical mechanisms. If the effects of all these mechanisms are described using a total wrinkling factor Ξ_{total}, then Eq. (3.29) can be modified as:

$$S_t = \Xi_{\text{total}} \cdot S_L. \tag{3.30}$$

The total wrinkling factor predicted by the multi-phenomena combustion model increases from 1.0 at the beginning to a maximum value of 1.689 near the end of the propagation. The growth of each wrinkling factor and the total wrinkling factor and their effect on combustion dynamics will be detailed in Chap. 4. It will be shown that the major contribution to the flame wrinkling comes from the turbulence generated by flame front itself, while the wrinkling due to the flow turbulence and the leading point mechanisms is very small. In order to investigate the influence of the wrinkling factors on the combustion dynamics, two additional LES simulations are performed on grid 1 using two different constant total wrinkling factors. Instead of Eq. 3.29 or 3.30 the turbulent burning velocity in these simulations is described by:

$$S_t = \Xi_{\text{const}} \cdot S_L, \tag{3.31}$$

where Ξ_{const} is the constant total wrinkling factor. The first additional simulation is conducted with the constant wrinkling factor $\Xi_{\text{const}} = 1.35$ (average between the minimum value 1.0 and the maximum average value of the wrinkling factor 1.689 in the simulations on grid 1 using the multi-phenomena combustion model) aiming to reproduce the average combustion rate during the whole combustion process. The second additional simulation is carried out with the constant wrinkling factor $\Xi_{\text{const}} = 1.08$ (average between the minimum value 1.0 and the average value of the wrinkling factor 1.153, at the moment the flame skirt touches the sidewalls in the original simulations with the multi-phenomena model), aiming to reproduce the combustion rate during the initial stage of flame propagation.

Figure 3.17 shows the most pronounced shape of both the tulip flame and the "distorted tulip" flame in a comparison between the experiment, the original LES simulation with the multi-phenomena combustion model, and the LES simulations with constant wrinkling factors $\Xi_{\text{const}} = 1.08$ and $\Xi_{\text{const}} = 1.35$ (the most pronounced tulip shape is established just before the tulip distortions). All of the three simulations reproduce the distortion of the tulip flame. However, all the distortions differ in details from the experimental distorted tulip flame which develops with clearly indented shape. Figure 3.17 shows comparison of the experimental pressure dynamics with the simulation results obtained for the multi-phenomena combustion model, and constant flame wrinkling factors of 1.08 and 1.35. It can be seen that the simulation with the multi-phenomena model reproduces experimental pressure transient with the best accuracy.

As it was expected the flame in the simulation with $\Xi_{\text{const}} = 1.08$ propagates nearly the same as that in the simulation with the multi-phenomena model in the early stages of combustion, and then gradually slower after the flame skirt touches the sidewalls. Consequently, the locations of both the most pronounced tulip and the "distorted tulip" flame are smaller in the simulation with $\Xi_{\text{const}} = 1.08$ than those in the multi-phenomena model as shown in Fig. 3.17, and the pressure also grows more slowly (see Fig. 3.18). However, a more pronounced tulip flame and "distorted tulip" flame can be created in the simulation with $\Xi_{\text{const}} = 1.08$

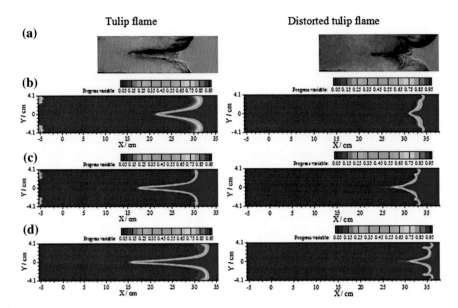

Fig. 3.17 The most pronounced shape of both tulip flame and "distorted tulip" flame in the experiment (**a**), the multi-phenomena model simulations (**b**), the simulations with $\Xi_{con} = 1.08$ (**c**) and simulations with $\Xi_{con} = 1.35$ (**d**)

Fig. 3.18 Pressure dynamics obtained in the experiment, simulation using the multi-phenomena model and simulations with constant wrinkling factors

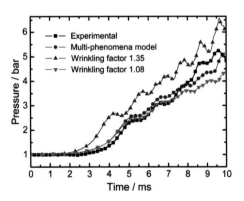

than the multi-phenomena model. In simulations with the flame wrinkling factor $\Xi_{const} = 1.35$ the flame at the initial stage of combustion propagates much faster, resulting in a faster pressure growth. In the later stages of combustion the pressure buildup rate is comparable to that in the experiment and for simulations with the multi-phenomena combustion model, though it oscillates more violently. Both the tulip flame and the "distorted tulip" flame are very pronounced in the simulation with $\Xi_{const} = 1.35$ that is thought due to higher amplitude of pressure oscillations.

Overall, the shape of the "distorted tulip" flame in simulations with the multi-phenomena combustion model and with the constant flame wrinkling factor $\Xi_{const} = 1.08$ generally resembles the experimental flame shape; in simulations with the flame wrinkling factor $\Xi_{const} = 1.35$ the "distorted tulip" shape is pronounced even better than in the simulations with the multi-phenomena combustion model. However, the simulations with the multi-phenomena combustion model delivered the best agreement with the experimental flame and pressure dynamics. This, in the authors' opinion, advocates the importance of SGS flame wrinkling modeling to reproduce correct flame acceleration in relatively large-scale explosions even in the absence of turbulence in the incoming flow.

3.4 Summary

The dynamics of premixed flame front propagation in the closed duct have been investigated using numerical simulations and compared to experiments. In the experiments, high-speed schlieren photography and pressure transducer were used. In the numerical simulations, the flame propagation was simulated as 2D and 3D chemically reacting flows. In order to overcome the difficulties arising from the wide range of length scales, either TF model or LES combustion model (burning velocity model) was used in the numerical simulations.

(1) Detailed experimental results were first shown for comparisons between numerical simulations and experiments. For the near-stoichiometric (hydrogen concentration 30 % by volume) hydrogen-air flame, the flame develops into a noticeable distorted tulip flame after a pronounced classical tulip flame has formed. Five stages of flame dynamics were proposed in the distorted tulip flame propagation, i.e., spherical flame, finger-shaped flame, flame with its skirt touching sidewalls, tulip flame, and distorted tulip flame. The distortions are initiated near the tips of the primary tulip lips. The distorted tulip flame develops into a salient "triple tulip" shape as the secondary tulip cusps approach the center of the primary tulip lips. The distorted tulip flame repeatedly undergoes collapse and reappearance processes. The first distorted tulip flame disappears as the secondary tulip cusps merge into the primary tulip cusp. Just before the disappearance of the primary distorted tulip flame, a second distorted tulip flame with a sequence of indentations created in succession on the primary tulip lips is formed. The second distorted tulip flame behaves in a similar way as the first one. Both the flame tip and pressure rise show periodic oscillations during the propagation of distorted tulip flame. The formation of the "distorted tulip" flame coincides with the sudden decrease in both the velocity of the leading flame front and the rate of pressure rise.

(2) In the 2D CFD calculations, a dynamically thickened flame model was employed to account for the premixed combustion. The dynamics of the premixed flame with distorted tulip shapes, observed in the experiments, has

been reasonably reproduced in the numerical simulation. The TF technique associated with the 19-step chemical reaction scheme is pretty reliable for studying the premixed hydrogen-air flame propagation and interaction between flame front and pressure waves in the closed vessel.

(3) In the 3D numerical simulations based on laminar solver, TF model was also used. Data obtained in the experiments, including high-speed schlieren images and pressure records, were used to elucidate the dynamics and evaluate the numerical model. The experimental results helped to assess the capability of the numerical model of reacting flow based on the solutions to the fluid flow equations and using thickened flame (TF) approach to simulate the premixed combustion process. The satisfactory agreement between the 3D numerical simulation and experiment indicates that the TF model with the seven-step chemistry scheme is reliable for predicting the transient premixed hydrogen-air combustion in the closed duct. The four stages of the classical tulip flame have been well reproduced in the numerical simulations. The numerical characteristic time for each stage is very close to those in the experiment and empirical model. The pressure dynamics has been also reasonably predicted in the numerical simulation, especially in the early stages of the flame propagation. Oscillations occur in the flame front dynamics and the pressure rise both in the experiments and numerical simulations after the flame reaches the lateral walls. Both the flame propagation speed and pressure buildup were underestimated after the formation of tulip flame in the numerical simulation. The flame front was artificially thickened in the numerical simulation to include several computational cells. As a result, the wrinkling of the flame front remains unresolved by the TF model in the present study.

(4) The multi-phenomena combustion model, a type of burning velocity model, has been applied in the LES simulations. The LES model reproduced the experimental observations with reasonable accuracy on the finer of the two grids applied. The good agreement between the numerical simulation and experiment confirms that the multi-phenomena combustion model provides a reasonable prediction of the premixed hydrogen-air combustion in the closed duct. The five typical stages of flame propagation, as observed in the experiment, are captured in the simulations run on the finer grid. The simulations on the course grid did not reproduce the "distorted tulip" flame. However, they did reproduce the "standard" tulip flame. The distortions, or smaller secondary cusps, are smoothed out on the courser grid due to the excessively thick numerical flame front, which is about 3–5 control volumes thick independent of control volume size when using the gradient method. The tulip flame in the simulations run on the coarse grid is less pronounced, as expected, and as a consequence the propagation velocity of the flame leading tip is underestimated to a larger extent when compared to the experimental results before the inversion of the flame front; the pressure buildup rate obtained on the coarse grid is also underestimated, but to a lesser extent.

References

1. Law CK (2007) Combustion at a crossroads: status and prospects. Proc Combust Inst 31:1–29
2. Wang F (2004) Computational fluid dynamics analysis: the principle and application of CFD software. Tsinghua University Press, Beijing (in Chinese)
3. Anderson JD (1995) Computational fluid dynamics: the basic with applications. McGraw-Hill, New York
4. Zhou X (1995) Computational hydraulics. Tsinghua University Press, Beijing (in Chinese)
5. Tao W (2001) Numerical heat transfer, 2nd edn. Xi'an Jiaotong University Press, Xi'an (in Chinese)
6. White FM (1991) Viscous fluid flow. McGraw-Hill, New York
7. Pope SB (2000) Turbulent flows. Cambridge University Press, Cambridge
8. Piller M, Nobile E, Thomas J (2002) DNS study of turbulent transport at low Prandtl numbers in a channel flow. J Fluid Mech 458:419–441
9. Petchenko A, Bychkov V, Akkerman V, Eriksson LE (2006) Violent folding of a flame front in a flame-acoustic resonance. Phys Rev Lett 97:164501
10. Janicka J, Sadiki A (2005) Large eddy simulation of turbulent combustion systems. Proc Combust Inst 30:537–547
11. Bychkov V, Akkerman V, Fru G, Petchenko A, Eriksson LE (2007) Flame acceleration in the early stages of burning in tubes. Combust Flame 150:263–276
12. Veynante D, Vervisch L (2002) Turbulent combustion modeling. Prog Energy Combust Sci 28:193–266
13. Ciccarelli G, Dorofeev S (2008) Flame acceleration and transition to detonation in ducts. Prog Energy Combust Sci 34:499–550
14. Makarov DV, Molkov VV (2004) Modeling and large eddy simulation of deflagration dynamics in a closed vessel. Combust Explos Shock Waves 40:136–144
15. Molkov V, Makarov D, Grigorash A (2004) Cellular structure of explosion flames: modeling and large-eddy simulation. Combust Sci Technol 176:851–865
16. Makarov D, Verbecke F, Molkov V, Roe O, Skotenne M, Kotchourko A, Lelyakin A, Yanez J, Hansen O, Middha P, Ledin S, Baraldi D, Heitsch M, Efimenko A, Gavrikov A (2009) An inter-comparison exercise on CFD model capabilities to predict a hydrogen explosion in a simulated vehicle refuelling environment. Int J Hydrogen Energy 34:2800–2814
17. Molkov V (2009) A multiphenomena turbulent burning velocity model for large eddy simulation of premixed combustion. In: Roy GD (ed) Nonequilibrium phenomena: plasma, combustion, atmosphere. Torus Press Ltd., Moscow. pp 315–323
18. Molkov V, Makarov D, Puttock J (2006) The nature and large eddy simulation of coherent deflagrations in a vented enclosure-atmosphere system. J Loss Prev Process Ind 19:121–129
19. Baraldi D, Kotchourko A, Lelyakin A, Yanez J, Gavrikov A, Efimenko A, Verbecke F, Makarov D, Molkov V, Teodorczyk A (2010) An inter-comparison exercise on CFD model capabilities to simulate hydrogen deflagrations with pressure relief vents. Int J Hydrogen Energy 35:12381–12390
20. Molkov V, Dobashi R, Suzuki M, Hirano T (2000) Venting of deflagrations: hydrocarbon-air and hydrogen-air systems. J Loss Prev Process Ind 13:397–409
21. Bychkov VV, Liberman MA (2000) Dynamics and stability of premixed flames. Phys Rep 325:116–237
22. Zeldovich YB, Barenblatt GI, Librovich VB, Makhviladze GM (1985) The mathematical theory of combustion and explosions. Consultants Bureau, New York
23. Molkov V, Makarov D, Schneider H (2006) LES modelling of an unconfined large-scale hydrogen-air deflagration. J Phys D Appl Phys 39:4366–4376
24. Poinsot T, Veynante D (2005) Theoretical and numerical combustion, 2nd edn. Edwards RT Inc, Philadelphia

25. Butler TD, O' Rourke PJ (1977) A numerical method for two-dimensional unsteady reacting flows. Proc Combust Inst 16:1503
26. Kuo KK (2005) Principles of combustion, 2nd edn. Wiley, New York
27. Legier JP, Poinsot T, Veynante D (2000) Dynamically thickened flame LES model for premixed and non-premixed turbulent combustion. In: Proceedings of the Summer Program
28. Colin O, Ducros F, Veynante D, Poinsot T (2000) A thickened flame model for large eddy simulations of turbulent premixed combustion. Phys Fluids 12:1843–1863
29. Lacaze G, Cuenot B, Poinsot T, Oschwald M (2009) Large eddy simulation of laser ignition and compressible reacting flow in a rocket-like configuration. Combust Flame 156:1166–1180
30. De A, Acharya S (2009) Large eddy simulation of a premixed Bunsen flame using a modified thickened-flame model at two Reynolds number. Combust Sci Technol 181:1231–1272
31. Petchenko A, Bychkov V (2004) Axisymmetric versus non-axisymmetric flames in cylindrical tubes. Combust Flame 136:429–439
32. Bi M (2001) A research on the pressure fields of unconfined flammable gas cloud explosions. Dalian University of Technology, Dalian (in Chinese)
33. Lamoureux N, Djebaili-Chaumeix N, Paillard CE (2003) Laminar flame velocity determination for H_2-air-He-CO_2 mixtures using the spherical bomb method. Exp Thermal Fluid Sci 27:385–393
34. Dahoe AE (2005) Laminar burning velocities of hydrogen-air mixtures from closed vessel gas explosions. J Loss Prev Process Ind 18:152–166
35. Tse SD, Zhu DL, Law CK (2000) Morphology and burning rates of expanding spherical flames in H_2/O_2/inert mixtures up to 60 atmospheres. Proc Combust Inst 28:1793–1800
36. Zimont VL, Lipatnikov AN (1995) A numerical model of premixed turbulent combustion of gases. Chem Phys Rep 14:993–1025
37. Babkin VS (2003) Personal communication. Institute of Chemical Kinetics and Combustion, Siberian Branch, Russian Academy of Science, Novosibirsk, Russia
38. Yakhot V (1988) Propagation velocity of premixed turbulent flames. Combust Sci Technol 60:191–214
39. Yakhot V, Orszag SA (1986) Renormalization-group analysis of turbulence. Phys Rev Lett 57:1722–1724
40. Pocheau A (1994) Scale-invariance in turbulent front propagation. Phys Rev E 49:1109–1122
41. Karlovitz B, Denniston DWJ, Wells FE (1951) Investigation of turbulent flames. J Chem Phys 19:541–547
42. Molkov VV, Nekrasov VP, Baratov AN, Lesnyak SA (1984) Turbulent gas combustion in a vented vessel. Combust Explos Shock Waves 20:149–153
43. Gostintsev YA, Istratov AG, Shulenin YV (1988) Self-similar propagation of a free turbulent flame in mixed gas mixtures combustion. Combust Explos Shock Waves 24:63–70
44. Molkov V, Verbecke F, Makarov D (2008) LES of hydrogen-air deflagrations in a 78.5-m tunnel. Combust Sci Technol 180:796–808
45. Kuznetsov VR, Sabel Nikov VA, Libby PA (1990) Turbulence and combustion. Hemisphere Publishing Corporation, New York
46. Conaire ÓM, Curran HJ, Simmie JM, Pitz WJ, Westbrook CK (2004) A comprehensive modeling study of hydrogen oxidation. Int J Chem Kinet 36:603–622
47. Li J, Zhao Z, Kazakov A, Dryer FL (2004) An updated comprehensive kinetic model of hydrogen combustion. Int J Chem Kinet 36:566–575
48. Miller JA, Mitchell RE, Smooke MD, Kee RJ (1982) Toward a comprehensive chemical kinetic mechanism for the oxidation of acetylene: Comparison of model predictions with results from flame and shock tube experiments. Proc Combust Inst 19:181–196
49. Gonzalez M (1996) Acoustic instability of a premixed flame propagating in a tube. Combust Flame 107:245–259
50. Clanet C, Searby G (1996) On the "tulip flame" phenomenon. Combust Flame 105:225–238
51. Matalon M (2009) Flame dynamics. Proc Combust Inst 32:57–82

52. Dunn-Rankin D, Sawyer RF (1998) Tulip flames: changes in shape of premixed flames propagating in closed tubes. Exp Fluids 24:130–140
53. Leyer JC, Manson N (1971) Development of vibratory flame propagation in short closed tubes and vessels. Proc Combust Inst 13:551–558

Chapter 4
Theoretical Analysis of Premixed Hydrogen–Air Flame Propagation in Ducts

4.1 Introduction

Theory of premixed flame propagation in tubes is an important subject in this area. The current theories may predict early laminar flame acceleration in a smooth tube. When the tube is obstructed, the flame can easily undergo transition to turbulent flame, even to detonation under proper conditions [1]. In addition, for the flame propagation in an enclosure, the interaction between flame and pressure waves play an important role. Autoignition and detonation can occur in the fresh mixture ahead of flame front due to the effects of strong pressure waves. These phenomena are unfavorable for safety and internal combustion engine design [2]. Although there have been a great number of studies of flame dynamics in confined regions [1–11], the understanding and theories are incomplete.

There are less theoretical investigations of premixed flame propagation in tubes compared to experimental and numerical studies, especially for the flame propagation with tulip and distorted tulip flames. As remarked above, the acceleration mechanism of a premixed flame in a half-open tube was first suggested and studied experimentally by Clanet and Searby [9]. They distinguished four stages of flame dynamics and proposed an empirical model to predict the characteristic time of each stage.

Following the pioneering work of Clanet and Searby [9] and Bychkov et al. [4] developed an analytical theory for the acceleration of a finger-shaped laminar flame and formation of the subsequent tulip flame in the early stages of flame propagation in long cylindrical half-open tubes, which is based on the assumption of an infinitesimally thin flame front. The model predicts growth rate, total acceleration time, maximum increase of the flame surface area, and intervals between the spherical, finger-shaped and tulip stages. Nevertheless, there may be still further work to be conducted to improve the theory. First, the theory does not describe the flame evolution after the initiation of tulip flame. Second, the theory is based on the assumption of an infinitesimally thin flame front. For a real flame, the finite flame

© Springer-Verlag Berlin Heidelberg 2016
H. Xiao, *Experimental and Numerical Study of Dynamics
of Premixed Hydrogen-Air Flames Propagating in Ducts*,
Springer Theses, DOI 10.1007/978-3-662-48379-4_4

thickness can not be zero. Furthermore, the effects of the flame intrinsic instabilities and state changes of unburned mixture are not taken into account. These effects can be significant when the flame accelerates fast and pressure waves interact with flame front. From the above experiments and numerical simulations, it has been known that a premixed hydrogen–air flame can experience further complex shape changes after formation of a pronounced classical tulip flame. Besides, the flame dynamics is also affected by various phenomena, such as gravity, opening ratio, and mixture composition.

4.2 Evolution of Premixed Flame in a Duct

A premixed flame propagating in a tube can undergo various shape changes, e.g., curved, planar, and cellular front [2, 12]. During these processes, the flame may accelerate or decelerate. Usually, a flame ignited at/near the closed end of a tube/duct by a point ignition source will develop semispherical/spherical, and then finger-like shapes. Figure 4.1 schematically shows the characteristics of early flame acceleration in a tube [4]. According to Clanet and Searby [9], the volume of burnt gas in the burning process can be expressed as:

$$\frac{dV_b}{dt} = E \cdot A_w \cdot S_L, \tag{4.1}$$

where V_b is the burnt gas volume, A_w is the total flame surface area. The expansion ratio $E = \rho_u / \rho_b$, where ρ_u and ρ_b are the densities of unburned and burnt gases, respectively. It is seen from Fig. 4.1 that the flame surface area comes mainly from the skirt of flame, so that the flame surface area can be estimated as $A_w \approx 2\pi H Z_{\text{tip}}$, where Z_{tip} is the position of the flame leading tip. Thus the volume of burnt gas can be approximately calculated as $V_b \approx \pi H^2 Z_{\text{tip}}$. Then Eq. (4.1) can be changed to:

$$\frac{dZ_{\text{tip}}}{dt} = 2E \cdot \frac{S_L}{H} \cdot Z_{\text{tip}}, \tag{4.2}$$

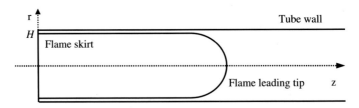

Fig. 4.1 Geometrical features of early flame acceleration in a tube. Reprinted from Ref. [4], Copyright 2007, with permission from Elsevier

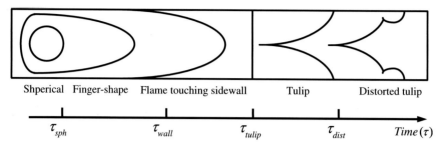

Shperical   Finger-shape   Flame touching sidewall   Tulip   Distorted tulip

τ_{sph} τ_{wall} τ_{tulip} τ_{dist} $Time\,(\tau)$

Fig. 4.2 Schematic showing typical flame evolution at five stages during hydrogen–air flame propagation in a closed duct

Following Eq. (4.2), the relationship between flame leading tip and time is given as:

$$Z_{tip} \propto \exp(2E \cdot \frac{S_L}{H} \cdot t). \tag{4.3}$$

The flame starts to accelerate as it evolves into finger-shaped stage. The flame acceleration terminates when the flame skirt reaches the tube sidewalls. According to the experiment of Clanet and Searby [9], this acceleration process lasts only a short time. As noted above, Bychkov et al. [4] suggested an analytical theory for calculating the characteristic time and maximum flame surface area at this stage.

Based on the experiments and numerical simulations presented above, the premixed hydrogen–air flame propagation in a closed tube comprises five stages and four characteristic times, as shown in Fig. 4.2. The five stages are spherical flame, finger-shaped flame, flame with its skirt touching tube sidewalls, tulip flame, and distorted tulip flame. The four characteristic times are the time duration of spherical flame τ_{sph}, the time when the flame touches the sidewalls τ_{wall}, the time when planar flame forms τ_{tulip}, and the time when distorted tulip flame is initiated τ_{dist}.

4.3 Factors Influencing the Flame Properties

Dynamics of a premixed flame in a tube depends on various factors which could be distinguished as internal and external parameters. The external factors include hydrodynamic instability, turbulence, vortex motion, pressure waves, wall friction, geometry features, etc. Internal factors can be fuel properties, composition and impurities of the reactive mixture, etc. The temperature and pressure of mixture can also be treated as internal parameter since they are always closely connected with flame properties. The internal parameters determine the basic flame properties while the external parameters may have different effects in different cases. Ignition energy directly determines the initial properties of a combustion wave.

Reactivity	Fuel
Table 4.1 Reactivity categories of common fuels	
Low	Ammonia, chlorethylene, methane
Medium	Ethylene, propane, ethane, n-butane
High	Acetylene, hydrogen, benzene

4.3.1 Influence of Fuel Properties

Combustion is substantially a rapid exothermic chemical process. The fuel properties have important effects on the combustion dynamics. From the point of view of thermodynamics, combustion is thermal activation of molecules. The lower the activation energy is, the faster the reaction is, and the higher the reaction rate is. Common fuels can be categorized into three types, depending on their reactivities, as shown in Table 4.1. As mentioned in Background and Introduction, hydrogen has large burning velocity and adiabatic flame temperature in air due to its high reactivity.

4.3.2 Influence of Mixture Composition

Composition of an energetic premixture can be described by equivalence ratio or concentration of fuel in the mixture. The mixture composition has significant influences on laminar burning velocity, adiabatic flame temperature, expansion ratio, and flame instabilities. The equivalence ratio for common hydrocarbon–air mixtures where the highest laminar burning velocity occurs is around $\Phi = 0.96$, independent of temperature and pressure. For hydrogen–air flame, the largest laminar burning velocity shifts to fuel-rich mixture with equivalence ratio $\Phi \approx 1.6$ [13]. Nevertheless, the stoichiometric hydrogen–air mixture has the highest temperature which is about 2400 K. The effects of equivalence ratio on premixed hydrogen–air flame propagation in ducts have been experimentally detailed in Chap. 2.

4.3.3 Influence of Pressure and Temperature

Pressure control plays an important role in engineering combustion. Usually, increasing pressure can lead to increase of combustion rate and efficiency. This is helpful in reducing the physical size of a combustion facility.

Generally, laminar burning velocity can be related to pressure as follows [2]:

$$S_L \propto p^n, \tag{4.4}$$

where n is a baric index. According to thermodynamic theory and ideal gas law $p \propto \rho/RT$, the dependence of laminar burning velocity on pressure can be given as:

$$S_L \propto p^{\frac{v}{2}-1}, \tag{4.5}$$

where v is the order of reaction. It is known from Eq. (4.5) that laminar burning velocity decreases as pressure increases with reaction order $v < 2$. When $v = 2$, laminar burning velocity is independent of pressure. When $v > 2$, laminar burning velocity increases with increasing pressure.

Following the Arrhenius law, increase of temperature of mixture results in increase of reaction rate and, therefore, increase of laminar burning velocity. The relationship between laminar burning velocity and temperature can be expressed as:

$$S_L \propto \exp(\frac{-E_a}{2RT_f}). \tag{4.6}$$

The effect of temperature of unburned mixture T_0 is included in the effect of flame temperature T_f on the laminar burning velocity. When the flame temperature is high enough (approximately > 2500 °C), the dependence of burning velocity on temperature may not be elucidated by thermodynamic theory.

The practical relationships between hydrogen–air laminar burning velocity and pressure and temperature can be described by Eqs. (1.6) and (1.8), respectively.

4.3.4   Influence of Impurities

When noncombustible or inert gases, e.g., Ar, N_2, CO_2, and He, are mixed with a reactive mixture, the physical and chemical properties of the mixture, such as heat conductivity and molecule diffusivity, would be changed. Thus the laminar flame speed would be changed as well. For example, the effects of CO_2 and N_2 on laminar burning velocity can be estimated as:

$$S_L = S_{L0} \cdot (1 - 0.01[N_2] - 0.012[CO_2]). \tag{4.7}$$

On the other hand, the laminar burning velocity increases as the oxidizer concentration increases. For example, the laminar burning velocity of fuel-oxygen mixture is much higher than that of fuel-air mixture.

4.3.5   Influence of Ignition Energy

Ignition energy is important for determining what type of combustion wave will be created. A laminar flame is usually initiated following a weak ignition while a detonation wave can be ignited by strong ignition. Besides, from the point of view of safety, lower ignition energy poses higher explosion risk. For example, natural gas has the lowest ignition energy of 0.28 mJ in air, so that it is very easy to ignite a natural gas flame in air [14]. This is one of the main reasons causing natural gas explosions in coal mines. The lowest ignition energy of hydrogen in air is even much lower, only 0.019 mJ.

4.4 Theoretical Analysis of Premixed Hydrogen–Air Flame in the Duct

4.4.1 Empirical Model

The experimental study by Kerampran et al. [15] shows that the increase of flame surface area is the only reason for flame acceleration in the early stage of premixed flame propagation in a tube. They found that the ratio of flame surface area to the area of tube cross-section is equal to the ratio of flame propagation speed to laminar burning velocity. The flame propagation speed in the axial direction increases rapidly with the fast increase in flame surface area as the flame evolves from spherical to finger-shaped shape. In the meantime, the flame propagation speed in the radial direction is close to laminar burning velocity since the expansion of combustion products near the sidewalls is prevented due to the confinement of tube sidewalls [16]. Clanet and Searby [9] proposed an empirical model for tulip flame propagation in a tube based on experiments. Following the geometrical assumption in Fig. 4.1, the position of flame leading tip can be calculated using an exponential expression:

$$\frac{Z_{tip}}{H} = \exp(\frac{t - t_{sph}}{\tau}), \tag{4.8}$$

where Z_{tip} is the location of flame leading tip defined as the distance to ignition point, $\tau = H/(2ES_L)$ is the characteristic time, and t_{sph} is the time when flame changes from spherical to finger-shaped shape, namely the time duration of spherical flame. The maximum flame propagation speed is reached as the flame starts to touch tube sidewalls. The time when the flame touches the sidewalls can be given by an empirical expression as [9]:

$$t_{wall} = (0.26 \pm 0.02) \cdot \frac{H}{S_L}. \tag{4.9}$$

The time t_{sph} can be calculated as follows using t_{wall} of Eq. (4.9) as the time variable in Eq. (4.8):

$$t_{sph} = t_{wall} - \frac{H}{S_L} \cdot \frac{1}{2E} \cdot \ln(\frac{Z_{wall}}{H}), \tag{4.10}$$

where Z_{wall} is the location of flame tip when flame skirt reaches the sidewalls of tube. The time t_{sph} can also be estimated by a linear relationship:

$$t_{sph} = (0.1 \pm 0.02) \cdot \frac{H}{S_L}. \tag{4.11}$$

In the third stage of flame propagation, the flame leading tip decelerates fast since the flame surface area reduces quickly. The propagation speed of flame skirt next to the tube sidewalls is nearly constant. A planar flame forms subsequent to the deceleration and then a tulip flame develops. The flame inversion time, i.e., the onset time of tulip flame, can be given as the time when the flame front curvature changes sign. Clanet and Searby [9] also empirically suggested a relationship of the formation time of tulip flame with the tube radius and laminar burning velocity:

$$t_{tulip} = (0.33 \pm 0.02) \cdot \frac{H}{S_L}. \tag{4.12}$$

It can be known from Eq. (4.11) and that $t_{tulip} \approx 1.29 t_{wall}$. After flame inversion, the flame deforms into a tulip flame.

4.4.2 Theoretical Model and Results

The acceleration mechanism of a premixed flame in a half-open tube was previously suggested and studied experimentally by Clanet and Searby [9]. Bychkov et al. [4] developed an analytical theory for the acceleration of a finger-shaped laminar flame and formation of the subsequent tulip flame in the early stages of flame propagation in long cylindrical half-open tubes, which is based on the assumption of an infinitesimally thin flame front. The model predicts growth rate, total acceleration time, maximum increase of the flame surface area and intervals between the spherical, finger-shaped, and tulip stages. In the theory, time scale and flame propagation speed are reduced by H, H/S_{L0} and S_{L0}, respectively. The flame is ignited at the center of the closed end and propagates toward the open one. According to the theory, the flame evolves from a spherical kernel to the finger-shaped front at a scaled time when the flame skirt has moved about halfway to the side wall:

$$\tau_{sph} = \frac{1}{2\alpha}, \tag{4.13}$$

where $\alpha = \sqrt{E \cdot (E - 1)}$. The flame acceleration starts as the flame develops from spherical to the finger-shaped front and stops when the flame skirt touches the walls. The scaled time at which the flame front touches the sidewalls is calculated as:

$$\tau_{\text{wall}} = \frac{1}{2\alpha} \cdot \ln\left(\frac{E + \alpha}{E - \alpha}\right). \tag{4.14}$$

The flame inversion happens at a time:

$$\tau_{\text{tulip}} = \tau_{\text{wall}} + \tau_{\text{inv}}, \tag{4.15}$$

where $\tau_{\text{inv}} = \lambda/\alpha$ is the time interval between τ_{wall} and τ_{tulip}. λ is a coefficient comparable to unity and was taken as $\lambda = 1$ in ref. [4]. The dimensionless position of the flame leading tip is given by:

$$\xi_{\text{tip}} = \frac{E}{4\alpha} \cdot [\exp(2\alpha\tau) - \exp(-2\alpha\tau)] = \frac{E}{2\alpha} \cdot \sinh(2\alpha\tau). \tag{4.16}$$

At the acceleration stage, the flame leading tip accelerates exponentially with a growth rate:

$$\theta = 2\alpha = 2\sqrt{E \cdot (E - 1)} \tag{4.17}$$

The flame surface area reaches its maximum value when the flame skirt starts to touch the sidewalls and the scaled maximum flame surface area is obtained by:

$$a_{\text{max}}^* = \frac{A_{\text{max}}}{\pi \cdot H^2} = 2\frac{E^2}{E + 1}, \tag{4.18}$$

where A_{max} is the physical flame surface area $(x > 0)$. Since the main contribution to the flame surface area is from the skirt of the flame front, then $A_{\text{max}} \approx 2\pi H \cdot Z_{\text{tip,wall}}$, where $Z_{\text{tip,wall}}$ is the position of the flame leading tip as the flame skirt touches the sidewalls.

The theory was validated against both the experiment by Clanet and Searby [9] and DNS with good agreement in [4]. Experimental data and simulation results obtained in the present study will be compared with the theory by Bychkov et al. [4], though it is worth noting that the theory is developed for flame propagation in long cylindrical half-open tubes, with nearly constant pressure ahead of flame front and constant (i.e., laminar) burning velocity. It is expected that the theory may be applicable in the early stages of flame propagation in the described experiment, when pressure changes are negligible and burning velocity is not affected by the developing instabilities. The analytical theory could lead to overestimation of the speed and position of flame tip when applied at the later stages of flame propagation in a closed tube: flame propagation will decelerate due to the inability of the gases to expand freely and flow toward the closed end. Besides, the skirt of the flame in a square cross-sectional duct may touch the sidewalls differently compared to that in a circular cross-sectional tube. On the other hand, the burning velocity in a closed tube is generally enhanced by the joint effects of wrinkling (turbulence) and compression. From this point of view, the increase of the burning velocity in a closed duct would compensate to some extent the propagation speed and position of the flame tip, making comparison of flame tip propagation dynamics with analytical theory at least complicated.

The theory by Bychkov et al. [4] can predict the tulip flame formation. However, currently there is no theory of distorted tulip flame propagation since it is a new flame phenomenon. From the experiments in Chap. 2 and the numerical calculations in Chap. 3, it is known that formation of distorted tulip flame is closed connected with pressure wave. Initiation of a distorted tulip flame is consistent with sudden flame deceleration induced by pressure wave. Therefore, the formation time and the corresponding flame tip position may be obtained by analyzing the interaction between flame front and pressure wave. Here, the sound speed is assumed to be a constant during distorted tulip flame formation. Dimensionless sound speed is given as $\varsigma = c_s/S_{L0}$, where c_s is the physical sound speed in hydrogen–air mixture. Pressure wave (acoustic wave) is triggered by the first contact of flame skirt with tube sidewalls. The contact time and flame tip position at the monument can be calculated using Eqs. (4.14) and (4.16), respectively. The pressure wave is reflected as it arrives at the right end wall. The reflected pressure wave passes through the flame front and causes the flame to decelerate. Then a distorted tulip flame is initiated subsequent to the flame deceleration. We consider the time when the reflected pressure wave meets the flame front as the initiation time of distorted tulip flame τ_{dist}, and the flame tip position at this time as $\xi_{tip,dist}$. The flame tip position after flame skirt touches tube sidewalls may be determined as [4]:

$$\xi_{tip} = -(\beta - 1) \cdot E \cdot \exp[\theta \cdot (\tau - \tau_{wall})] + \beta \cdot E \cdot [\theta \cdot (\tau - \tau_{wall})], \quad (4.19)$$

where $\beta = 1.25$ is a model constant. Thus the flame tip position when distorted tulip flame starts to form might be given as:

$$\xi_{tip,dist} = -(\beta - 1) \cdot E \cdot \exp[\theta \cdot (\tau_{dist} - \tau_{wall})] + \beta \cdot E \cdot [\theta \cdot (\tau_{dist} - \tau_{wall})]. \quad (4.20)$$

And the initiation time of distorted tulip flame can be obtained as:

$$\tau_{dist} = (L/H - \xi_{tip,wall})/\varsigma + (L/H - \xi_{tip,dist})/\varsigma, \quad (4.21)$$

where L is the length of tube.

Finally, the initiation time of distorted tulip flame τ_{dist} and the corresponding flame tip $\xi_{tip,dist}$ position may be estimated using Eqs. (4.14), (4.20) and (4.21).

4.5   Comparisons Between Experiments, Numerical Simulations and Theoretical Predictions, and the Combustion Regime

Figure 4.3 shows the position (a) and propagation speed (b) of the flame leading tip in a comparison between experiment, theory and LES simulation (Grid 1). For comparison with the theory the length scale, time scale, and flame propagation speed in the experiment and numerical simulation are also scaled by H, H/S_{L0} and S_{L0}, respectively. For further comparison, the growth rate θ, maximum reduced flame surface area a^*_{max} and the characteristic reduced times of spherical τ_{sph}, finger-shaped τ_{wall} and tulip τ_{tulip} stages are summarized in Table 4.2.

As mentioned in Sect. 4.4.2, the theory should overestimate the speed and position of the flame tip when applied to flame propagation in the closed duct.

Fig. 4.3 Location (**a**) and propagation speed (**b**) of the flame leading tip in the experiment, theory, and LES simulation (Grid 1)

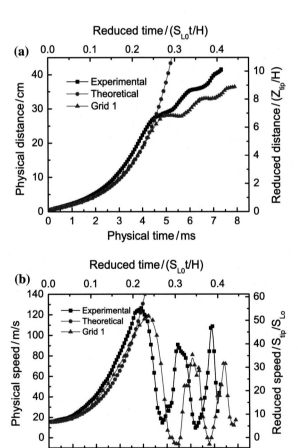

Table 4.2 Parameters for the experimental, numerical and theoretical comparisons

	τ_{sph}	τ_{wall}	τ_{tulip}	τ_{dist}	θ	a^*_{max}
Experimental	0.060	0.205	0.272	0.335	14.40	10.062
Theoretical	0.079	0.256	0.336	–	12.60	11.896
Grid 1	0.071	0.223	0.296	0.384	12.782	10.585
Grid 2	0.072	0.218	0.302	–	12.576	9.112

Fig. 4.4 The development of averaged SGS wrinkling factors through the flame front

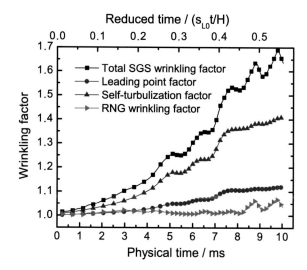

Nevertheless, it can be seen from Fig. 4.3a that the flame tip position for the LES simulations and the theory coincide with each other at the early spherical and finger-shape stages (the position of the flame tip in simulation on grid 1 is slightly larger than the theoretical value). On the contrary, the flame speed in the LES simulations increases even faster than in the theory at the acceleration stage, as shown in Fig. 4.3b. This behavior is due to flame acceleration mechanisms described in Sect. 3.2.3.2 which play a role in this relatively large-scale experiment. The development of the averaged (through the flame front) SGS wrinkling factors is shown in Fig. 4.4 (grid 1).

The averaging of the SGS wrinkling factors in the flame front ($0.01 \leq c \leq 0.99$) is obtained by:

$$\Xi_{ave} = \frac{\int_{0.01}^{0.99} \Xi \cdot |\nabla \tilde{c}| dV}{\int_{0.01}^{0.99} |\nabla \tilde{c}| dV}, \tag{4.22}$$

where Ξ is a wrinkling factor: either the leading point wrinkling factor Ξ_{lp}, or the flame self-turbulization factor Ξ_K, or the factor Ξ_{RNG} representing the effect of flow turbulence in the unburned gas ($\exp{(u'/S_t)^2}$). The total SGS wrinkling factor is calculated as $\Xi_{total} = \Xi_{lp} \cdot \Xi_K \cdot \Xi_{RNG}$. The flame acceleration caused by the flow turbulence (RNG wrinkling factor) is very small during the whole process of the combustion, reaching a value of about 1.08 only at the end of combustion process, which implies that the flow is nearly laminar or weakly wrinkled. On the other hand, the total SGS flame wrinkling factor reaches a value of 1.689, which shows the importance of SGS modeling of other flame wrinkling factors for the considered experiment.

The total SGS wrinkling factor increases from 1.0 to 1.153 until the skirt touches the sidewalls at $t = 4.0679$ ms (grid 1), as shown in Fig. 4.4, and pressure in the duct grows to $p/p_0 = 1.594$. With laminar burning velocity increase due to compression of the unburned mixture as much as $(p/p_0)^{0.52} = 1.274$, the increase of the turbulent burning velocity compared to the initial laminar one S_t/S_{L0} reaches a factor of 1.469 at the end of the flame acceleration. The theory [4] does not account for this increase in the burning rate of almost 50 %. Fortunately, the flame acceleration stage in the finite-size duct is short in time and the difference of the flame speed does not lead to a significantly different flame tip position (for a larger scale experiment this discrepancy would be larger). As it is shown above, pressure built up during this initial stage of the flame propagation is relatively small, expansion of the combustion products is almost unaffected by the pressure rise in the finite-size duct, and the results of experiment, theory, and LES simulations for the time of flame front transition from spherical to finger-shape τ_{sph} are very close, see Table 4.2. The characteristic times of finger-shape (τ_{wall}) and tulip (τ_{tulip}) stages in the theory are larger than those in the experiment and LES simulations, which confirm the importance of the flame acceleration in the considered experiment. However, the maximum flame surface area in the theory is larger than those in the experiment and LES simulations. It can be concluded from Fig. 4.3 and Table 4.2 that at the early stage of flame development (spherical flame propagation and, partially, finger shaped) both the LES simulations and the theory reproduce the experiment reasonably well, but at the later stages of finger-shaped and tulip formation the LES simulations, accounting for flame acceleration and 3D-closed tube geometry, are in a better agreement with the experiment.

4.6 Effects of Wall Friction

In order to examine the effects of wall friction on premixed hydrogen–air flame, the dynamically thickened flame model described in Sect. 3.2.3 is used to simulate the flame propagation. For comparisons, two different boundary conditions are employed at the walls of the duct in the numerical simulations: (1) nonslip and adiabatic boundary conditions and (2) free-slip and adiabatic boundary conditions. The initial conditions are the same as those in the experiment. The flammable mixture is ignited by a prescribed circular plot (with a radius of 2 mm) of hot gas. The gravity is neglected since the Froude number $F_r = S_L^2/(gH) \approx 9.2$, where g = 9.8 m/s^2 and $H = 4.1$ cm are the gravity acceleration and half of the height of the duct, respectively.

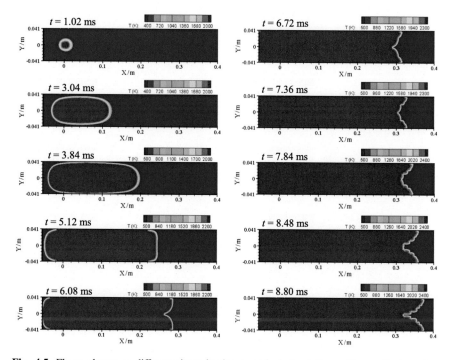

Fig. 4.5 Flame shapes at different times in the numerical simulation with nonslip boundary condition at the walls of the duct. The colors describe the temperature from unburned mixture (*blue*) to burnt gas (*red*)

Figure 4.5 shows the time history of flame evolution in the numerical simulation with nonslip boundary condition at the duct walls (nonslip simulation). The five stages of flame dynamics observed in the experiments are reasonably reproduced in the numerical simulation. Similar to that in the experiment, the flame experiences a short spherical stage after ignition and the finger flame starts at about $t = 1.02$ ms. The flame touches the lateral walls of the duct at about $t = 3.84$ ms, which is slightly later than that in the experiment (3.533 ms). The flame front becomes planar at $t = 5.12$ ms and a tulip-shaped front is formed. The distortions are initiated after $t = 6.08$ ms. Thereafter, a noticeable distorted tulip shape is produced (e.g., the flame shape at $t = 7.36$ ms in Fig. 4.5). The distorted tulip flame vanishes with the distortions and the primary tulip cusp propagating one to the other. The tulip flame is recovered after the collapse of the distorted tulip flame and appears quite close to the experimental observation.

Figure 4.6 presents the flame evolution in the later stages in the numerical simulation with free-slip boundary condition at the walls of the chamber (free-slip simulation). The flame development with free-slip boundary condition in the early stages is almost the same as that with nonslip boundary condition. After the inversion a tulip flame forms and behaves in the same way as that in the nonslip simulation. A distorted tulip flame is also generated after the formation of a

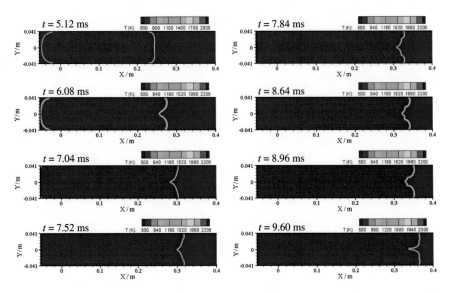

Fig. 4.6 Flame front evolution with time in the later stages in the numerical simulation with free-slip boundary condition at the duct walls

well-pronounced tulip flame, as seen in Fig. 4.6. Much like that in the nonslip simulation, the distortions propagates toward the primary tulip cusp and as they approach the center of the primary tulip lips a remarkable distorted tulip flame is formed, e.g., the flame at $t = 7.84$ ms in Fig. 4.6. In addition, the distorted tulip flame collapses with the distortions and the primary tulip cusp meeting each other and a classical tulip flame is repeated (see the flame shape at $t = 9.6$ ms in Fig. 4.6). Therefore, it can be concluded that both the tulip and distorted tulip flames can be formed in the absence of wall friction. Nevertheless, the numerical simulation with free-slip boundary condition show different characteristic features compared to that with nonslip boundary condition. First, both the onset time ($t = 7.04$ ms) of the distortions and the corresponding location ($z = 30.5$ cm, where d is the distance to the ignition site) of flame tip near the duct sidewalls in the free-slip simulation are obviously larger than those in the nonslip simulation ($t = 6.08$ ms and $z = 28.6$ cm). Second, the flame tip close to the sidewalls in the free-slip simulation travels at a relatively smaller speed at the later flame stage than that in the nonlip simulation and the second tulip flame is significantly less pronounced.

Figures 4.7 and 4.8 present the location and speed of flame leading tip with time in the experiments, numerical simulations with nonslip and free-slip boundary conditions at the walls of the duct, respectively. The location is the distance from the ignition site. Before the tulip formation the flame front along the centerline of the duct is the leading tip. After the inversion, the flame tip next to the upper wall is taken as the leading tip. The sudden pressure drop at around $t = 10$ ms corresponds to the opening of a vent which is set up near the right endwall of the duct to ensure safety in the experiments. The flame accelerates fast before the arrival of the flame

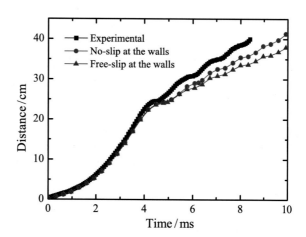

Fig. 4.7 Location of flame leading tip as a function of time in the experiment, numerical simulations with nonslip and free-slip boundary conditions

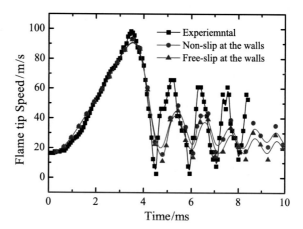

Fig. 4.8 Propagation speed of flame leading tip versus time in the experiment, numerical simulations with nonslip and free-slip boundary conditions

skirt at the sidewalls due to the exponential increase of the flame surface at the second stage. At the third stage, the flame decelerates quickly due to the severe reduction in the flame surface area. At the fourth stage, the flame accelerates again with the development of the tulip shape which leads the flame surface area to grow.

The second and the later flame decelerations are caused by the interaction of the flame front with the pressure wave, as will be shown in Chap. 5. The pressure wave drives the flame to display a periodically oscillating behavior. The pressure wave is triggered by the first contact of the flame with the sidewalls of the duct and moves forth and back inside the vessel. When the pressure wave reaches the right endwall it will be reflected. The passage of the pressure wave through the flame front results in a sudden flame deceleration. The formation of the tulip and distorted tulip flames coincides with the sudden decrease of both the flame displacement speed (propagation speed) and pressure growth rate inside the chamber. The oscillating period derived from the pressure trajectory in the experiment is about 1.18 ms. And the period in the numerical simulations is around 1.17 ms, very close to the experimental one. It can be seen that the period and amplitude of the flame oscillations in free-slip simulation is nearly the same as that in the nonslip simulation. This indicates that the mechanism of flame-pressure wave interaction remains unchanged with the variation of the boundary condition. A tulip flame is produced subsequent to the first flame deceleration in the experiment, nonslip and free-slip simulations. The distorted tulip flame is initiated at $t = 6.08$ ms (see Fig. 4.5) in the nonslip simulation immediately after the flame achieves a second deceleration. However, in the free-slip simulation, the flame does not assume a distorted tulip shape during the second deceleration, instead, the distorted tulip shape starts to form at about $t = 7.04$ ms till the flame undergoes a third deceleration. Another discrepancy is that the movement of the flame front is faster after the contact with the duct sidewalls when wall friction is taken into account, as shown in Figs. 4.7 and 4.8. This difference is also presented in Fig. 4.9, which shows the pressure dynamics as function of time for the experiment, nonslip and free-slip simulations.

The typical flow fields near the flame front at $t = 6.4$ ms in the nonslip (a) and free-slip (b) simulations are described in Fig. 4.10. The free-slip boundary condition at wall means that the longitudinal velocity components in both the unburned and burnt gas can freely acquire any value across the flame front and next to the duct wall. In contrast, the nonslip condition leads the longitudinal velocity components to decrease to zero when approaching the duct wall. In order to satisfy the continuity equation across the flame and next to the wall, the flame front assumes an angle with the wall of the duct. Consequently, a higher longitudinal flame propagation speed near the wall is caused. This result agrees with the analysis of Marra and Continillo [17] on the interaction of flame with wall. Besides, a more pronounced tulip flame is repeated in the presence of wall friction after the collapse of the distorted tulip flame, as shown in Fig. 4.5.

In the early stages, the flame propagation speed and pressure build-up in the numerical simulations are in close agreement with those in the experiment. Starting

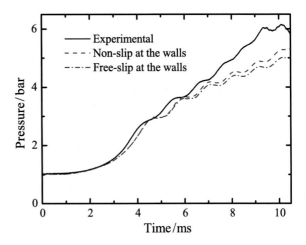

Fig. 4.9 Temporal evolution of pressure inside the combustion chamber

Fig. 4.10 Comparison of flow field around the flame front in the simulations with nonslip (**a**) and free-slip boundary conditions (**b**) at the chamber walls at $t = 6.4$ ms. Wall at the bottom in both pictures. The contour lines (*red*) present the amplitude of the longitudinal velocity component. The arrow lines (*black*) describe the direction and amplitude of the vector velocity

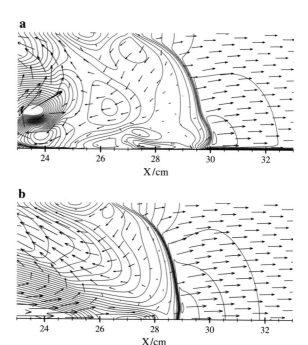

from the third stage, the numerical flame advances slower and the discrepancy becomes larger with time. This is thought to be due to the thickened flame front. The effect of the flame wrinkling can not be resolved in the numerical simulations using a TF technique since the flame is generally wrinkled at the scales below the grid resolution adopted in the present simulations. Overall, though a simplified numerical method, i.e., a 2D laminar approach with a dynamically thickened flame model, is applied in this study, the numerical simulations correctly predict the essential characteristic features of the flame observed in the experiments, including the flame shape changes, flame movement speed, pressure growth, and periodically oscillating behavior.

4.7 Summary

This chapter has described a theoretical analysis of premixed hydrogen–air flame propagation in a duct. The results were compared with experimental and LES results. An analytical model was suggested for the flame propagation with distorted tulip shape. The combustion regime was clarified using LES simulations. In addition, the effects of wall friction were also examined.

(1) Based on the previous analytical studies by other researchers, experiments, and LES simulations in this work, the five stages of dynamics of distorted tulip flame were examined in detail. A theoretical model was preliminarily suggested by analyzing the interaction between flame front and pressure waves. Agreement between the experiment, theory, and LES simulations at the early stage of the flame propagation is satisfactory.

(2) It was shown by the LES simulations that SGS wrinkling effects are important for flame acceleration while the turbulence in the incoming flow plays a minor role in the premixed hydrogen–air flame propagation in the duct. The major wrinkling effect comes from the self-turbulization of flame front. The flow was found to be nearly laminar, but modeling of the various mechanisms that enhance the burning rate is necessary.

(3) It was demonstrated that both the tulip and distorted tulip flames can be produced in the absence of wall friction. This means that wall friction may be not important for the formation of both tulip and distorted tulip flames in the duct considered in the present work. The change of boundary conditions at walls cannot substantially change the interaction of flame front with pressure waves. Nevertheless, the boundary layer has an effect to a certain extant on the flame propagation near the sidewalls of the vessel, leading to faster flame movement and higher pressure growth rate after the flame reaches the sidewalls. The flame leading tip moves at a higher speed and a more pronounced tulip shape is repeated subsequent to the vanishing of the distorted tulip flame in the nonslip numerical simulation than in the free-slip simulation.

References

1. Ciccarelli G, Dorofeev S (2008) Flame acceleration and transition to detonation in ducts. Prog Energy Combust Sci 34:499–550
2. Bychkov VV, Liberman MA (2000) Dynamics and stability of premixed flames. Phys Rep 325:116–237
3. Markstein GH (1964) Nonsteady flame propagation. Pergamon Press Limited, New York
4. Bychkov V, Akkerman V, Fru G, Petchenko A, Eriksson LE (2007) Flame acceleration in the early stages of burning in tubes. Combust Flame 150:263–276
5. Dunn-Rankin D, Barr PK, Sawyer RF (1986) Numerical and experimental study of "tulip" flame formation in a closed vessel[J]. Proc Combust Inst 21:1291–1301
6. Dunn-Rankin D, Sawyer RF (1998) Tulip flames: changes in shape of premixed flames propagating in closed tubes. Exp Fluids 24:130–140
7. Dorofeev S (2008) Flame acceleration and transition to detonation: a framework for estimating potential explosion hazards in hydrogen mixtures. Lecture presented at the 3rd European Summer School on Hydrogen Safety. Belfast, UK
8. Gonzalez M, Borghi R, Saouab A (1992) Interaction of a flame front with its self-generated flow in an enclosure—the tulip flame phenomenon. Combust Flame 88:201–220
9. Clanet C, Searby G (1996) On the "tulip flame" phenomenon. Combust Flame 105:225–238
10. Gamezo VN, Ogawa T, Oran ES (2007) Numerical simulations of flame propagation and DDT in obstructed channels filled with hydrogen-air mixture. Combust Flame 31:2463–2471

11. Dorofeev SB (2002) Flame acceleration and DDT in gas explosions. J Phys 12:3–10
12. Gonzalez M (1996) Acoustic instability of a premixed flame propagating in a tube [J]. Combust Flame 107:245–259
13. Law CK (2006) Combustion physics. Cambridge University Press, New York
14. Wu H, Zhang L, Guo Z (2004) Experimental research of igniting energy for the methane flame propagation. Coal Mine Blasting 1:5–7 (in Chinese)
15. Kerampran S, Desbordes D, Veyssière B (2001) Propagation of a fame from the closed end of a smooth horizontal tube of variable length. In: proceedings of the Proceedings of the 18th ICDERS, F
16. Guenoche H (1964) Flame propagation in tubes and in closed vessels. In: Markstein GH (ed) Nonsteady flame propagation. Pergamon Press, New York, p 107
17. Marra FS, Continillo G (1996) Numerical study of premixed laminar flame propagation in a closed tube with a full Navier-Stokes approach. Proc Combust Inst 26:907–913

Chapter 5
Mechanisms of Flame Deformations in the Premixed Hydrogen–Air Flame Propagation

5.1 Introduction

From above analysis, it is known that a premixed hydrogen–air flame in a duct can undergo a series of continuous shape changes, namely spherical flame, finger-shaped flame, flame with its skirt touching sidewalls, tulip flame, and distorted tulip flames. Here emphasis is put on the formation mechanisms of tulip and distorted tulip flames since these two curious flame phenomena are extremely complex and there is a lack of conclusive explanations for them.

The study of tulip flame began in the early twentieth century. Photographs of premixed flames propagating in closed vessels were first reported by Ellis [1] in 1928, who found that the flame shape spontaneously undergoes an inversion: it changes suddenly from a forward pointing finger to a backward pointing cusp in closed tubes under specific geometrical conditions, i.e., with an aspect ratio larger than two. This flame shape was subsequently referred to as "tulip" flame [2]. However, the notion of "tulip" flame could be misleading [3], since the "tulip" shape may be generated due to many different mechanisms with little or no relation between each other, e.g., Darrieus–Landau (DL) instability [4], flame-shock interaction [5], and deceleration of finger-shape flame front [1, 6–10]. The "tulip" flame considered in the present work is the sudden flame inversion during the deceleration of a finger-shape flame. Another curious observation is that in very long tubes (with aspect ratio about 20), the inversion of the flame front can reverse itself [11]. The flame shape becomes again convex toward the unburned mixture. This process repeats itself several times with a series of flame front inversions until the flame reaches the end of the tube.

There have been a large number of experimental, analytical, and numerical investigations of the tulip flame phenomenon aiming at explaining the mechanisms responsible for its formation [3, 6–10, 12–16]. Various possible explanations have been put forward such as: quenching and viscosity effects [1, 7, 17], the interaction between flame and pressure wave [11], burnt gases vortex motion effect [6, 18–20],

© Springer-Verlag Berlin Heidelberg 2016
H. Xiao, *Experimental and Numerical Study of Dynamics of Premixed Hydrogen-Air Flames Propagating in Ducts*, Springer Theses, DOI 10.1007/978-3-662-48379-4_5

the DL flame instability [10, 12, 15, 16], and Taylor instability [13]. Dunn-Rankin et al. [8], Rotman and Oppenheim [14], and Marra and Continillo [17] performed numerical simulations using zero Mach number model and demonstrated that the tulip flame can be formed in the absence of pressure wave effects. Clanet and Searby [13] concluded that neither boundary layer nor acoustic effects are dominant in tulip flame development. Gonzalez et al. [10] suggested that wall friction is unimportant for tulip flame formation. Marra and Continillo [17] disagreed with this conclusion and argued that wall friction is the determining cause for the onset of a tulip shape from a flat shape. Bychkov et al. [3] developed an analytical theory of the acceleration of a finger-shaped flame and the formation of a tulip flame in the early stages of laminar flame propagation in long cylindrical half-open tubes. The theory suggests that both the acceleration of the finger-shaped flame and formation of the tulip flame do not depend on the Reynolds number. It appears that currently there is no decisive and single explanation of the tulip flame formation mechanism. The understanding of tulip flame formation is incomplete and we will discuss it further in the present work.

As remarked earlier, a new flame phenomenon, distorted tulip flame, has been found in the experiments of hydrogen–air flame propagation in the present work. A distorted tulip flame is formed after a well-pronounced classical tulip flame has been established. The high-speed schlieren images showed that the primary lips of classical tulip flame are concaved toward duct sidewalls, producing noticeable secondary cusps. The dynamics of distorted tulip flame differs from that of classical tulip flame. In addition, as demonstrated in Chaps. 2 and 3, distorted tulip flame is also different from "double tulip" [10] and "multi tulip-shaped" [21] flames. Distorted tulip flame is a newly reported finding, and so far there have been few explanations on its formation. One of the main objectives of this study is to gain an insight into the physical mechanisms that lead to the formation of distorted tulip flame.

5.2   Interactions Between Flame and Pressure Waves

The experiments and numerical simulations used in this section are the same as those in Sect. 3.3.1.2, respectively. Figure 5.1 presents the experimental and numerical displacement speed (propagation speed) of the flame leading tip with time. The flame propagation speed oscillates periodically during its propagation both in the experiment and numerical simulation. The onset of the tulip and distorted tulip shapes coincides with the sudden flame deceleration. The first deceleration during the tulip flame formation results from the rapid decrease of the flame surface area after the flame has touched the sidewalls. However, the mechanism causing the second and the later decelerations is different. In fact, it is found that the distorted tulip flame is formed without remarkable reduction of the flame surface area, as shown in Fig. 3.5 ($t = 7.0$ ms).

Fig. 5.1 Experimental and numerical propagation speed of the flame leading tip

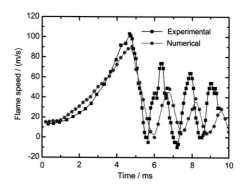

The flame deceleration is more drastic in the experiment, and the experimental flame leading tip propagates backward for a short while in the formation process of the distorted tulip flame. This implies that a stronger decrease of the flow velocity in the unburned gas is caused in the experiment. Note that the flame travels at a lower speed at the later stage in the numerical simulation than in the experiment. This also results from the thickened flame front in the numerical simulation. Actually, flame wrinkles grow over the flame front in the flame propagation and the flame has been obviously wrinkled before the flame inversion, as shown in Fig. 3.5. The appearance of flame wrinkles leads to an increase in the flame surface area and consequently an augmentation in the flame speed [22]. Nevertheless, the flame is generally wrinkled at the scales below the grid resolution considered in the current numerical simulation.

Figure 5.2 shows the experimental and simulated pressure growth rate derived from the pressure data recorded at the same location (at the bottom of the duct 400 mm from the ignition point in x-direction). At the initial stage, the flame is unaffected by the lateral walls of the duct and the pressure does not increase. The exponential increase of the flame surface area at the finger-shape stage leads the pressure to grow fast. The pressure growth rate drops quickly with a main part of the flame front quenching near the duct sidewalls after $t = 4.667$ ms in the experiment and $t = 4.8$ ms in the numerical simulation, respectively. A pressure

Fig. 5.2 Pressure growth rate in the experiment and numerical simulation

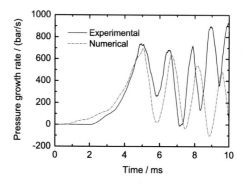

wave is initiated by the first contact of flame with the sidewalls and moves forth and back in the combustion vessel.

The pressure wave is reflected as it reaches the right endwall of the vessel. The pressure growth rates increases again with the formation of the tulip flame. Oscillations of pressure growth gate are triggered by the sudden flame deceleration at the third stage. The pressure growth rate oscillates periodically in phase with the flame tip propagation speed (Fig. 5.1). Both the tulip and distorted tulip flames are initiated after the pressure growth undergoes a steep deceleration. The time interval between the first and the second distorted tulip flames in the experiment is approximately 1.33 ms. The experimental period of the pressure wave obtained from the pressure growth rate is about 1.30 ms, nearly equal to the time interval between the first and the second distorted tulip flames. The situation in the numerical simulations with the period of pressure wave 1.40 ms is very similar to that in the experiment. Note that the numerical pressure growth rate is slightly larger than that in the experiment at the early stage. However, the pressure grows faster at the later stage in the experiment than in the numerical simulation due to the wrinkling effect.

Gonzalez [23] numerically found that the flame dynamics at the later stage in a closed vessel is in connection with the pressure wave. Before the onset of the distorted tulip flame, the interaction of the flame with pressure wave is commonly very weak. The contact of the flame front with the sidewalls leads to the first drastic flame deceleration and subsequently a pressure wave. When the pressure maximum passes the flame front in the direction from unburned mixture to burnt gas, the flame undergoes another sudden deceleration (see Fig. 5.1) and the pressure growth rate in the unburned region decreases violently again (see Fig. 5.2). The pressure growth rate reaches its second minimum value at $t = 6.86$ ms in the experiment and $t = 7.46$ ms in the numerical simulation, respectively, just before the initiation of the distortions both in the experiment ($t = 7.0$ ms) and the numerical simulation ($t = 7.6$ ms).

Figure 5.3 displays the evolution of the longitudinal velocity of the unburned gas versus time at a point located on the duct axis 400 mm from the ignition point in the numerical simulation. The velocity oscillations are also initiated by contact of the

Fig. 5.3 Numerical time evolution of the longitudinal component of velocity at the location on the symmetry axis 400 mm from the ignition point

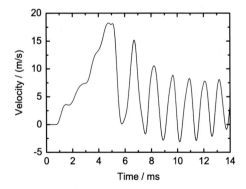

flame front with the duct sidewalls. The time period is the same as that of the pressure oscillations. The amplitude of the velocity oscillations decreases as the flame front approaches the right end wall of the duct.

The evolution of the flame shape (see Sect. 3.3.1.2), flame displacement speed, pressure growth rate, and flow velocity in the unburned mixture reveals that the interaction between the distorted tulip flame, and the pressure wave is quite strong. It has been numerically put into evidence by Gonzalez [23] that the oscillating behavior of a premixed flame in a closed vessel at the later stage is caused by the interaction of the flame with pressure waves. In his study, the periodic flame deceleration leads the flame front to exhibit an oscillating cellular pattern, which is akin to Taylor instability. In the experiments by Markstein [5], a convex flame is concaved in the center after the passage of a shock wave in the direction from the fresh mixture (high density) to the burnt region (low density). Markstein proposed that this behavior results from the violent velocity decrease behind the shock and explained it with the help of Taylor instability. The present numerical simulation gains further insight into physical process of the distorted tulip formation. A drastic decrease of the flow velocity in the unburned region with the initiation of the distorted tulip shape is produced due to the effect of the pressure wave, causing stagnant and even reverse flow in the unburned mixture, as shown in Fig. 5.3. The resulting velocity changes are comparable with those in the experiments of Markstein [5]. The onset of the distortions is accompanied by a sudden flame deceleration, and subsequently two secondary tulip cusps are created on the primary tulip lips. Therefore, the present study corroborates the conclusion that the distorted tulip flame has the same physical origin as that in the Markstein's experiments [5]. In addition, the fact that the distorted tulip flame is not formed in a half-open tube with the flame propagating from the closed end to the open one (see Chap. 2) confirms the importance of the pressure wave effect in the formation of a distorted tulip flame.

5.3   Formation Mechanism of Tulip Flame—Interactions of Flame with Flow

Figure 5.4 displays the vector velocity fields and vortex motion in the numerical simulation at different time instants $t = 8.0$ ms (a), 9.6 ms (b), 11.0 ms (c), and 16.6 ms (d). The flow opposite to the direction of the flame propagation is defined as reverse flow. The flow near flame front in burnt region is substantially vertical due to the presence of baroclinic effect [6, 24]. Navier–Stokes can be expressed as a vorticity equation:

$$\frac{d\vec{\Omega}}{dt} - (\vec{\Omega} \cdot \nabla \vec{v}) + \vec{\Omega}(\nabla \cdot \vec{v}) = -\nabla \times \left(\frac{\nabla p}{\rho}\right), \tag{5.1}$$

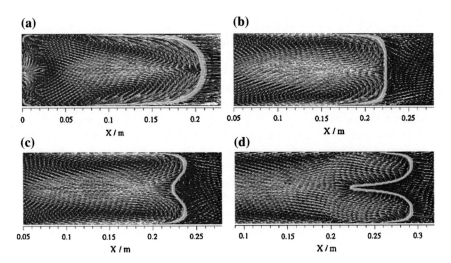

Fig. 5.4 Numerical vector flow velocity fields and flame shapes at the time instants $t = 8.0$ ms (**a**), 9.6 ms (**b**), 11.0 ms (**c**), and 16.6 ms (**d**)

where $\vec{\Omega} = \nabla \times \vec{v}$ is the vorticity. The baroclinic term of the equation can be given as:

$$-\nabla \times \left(\frac{\nabla p}{\rho}\right) = \frac{\nabla p \times \nabla \rho}{\rho^2}. \tag{5.2}$$

The baroclinic term indicates that if there is a misalignment between density gradient and pressure gradient, vorticity is produced due to the baroclinic effect. This implies that when a curved flame front is exposed to pressure waves, vorticity will be generated in the vicinity of flame front. Large-scale vortex can be created by the accumulation of vorticity. The flame shape would be changed due to the advection effects of vortical motion of flow [6, 19].

The present numerical results reveal that the reverse flow is generated near the middle section in the burned gas prior to the two symmetrical vortices which start to appear at the left closed end. The reverse flow becomes stronger as the flame continues to propagate. Whereafter, two small vortices are initiated near the two sidewalls due to the effects of the reverse flow, confinement of the two sidewalls and left closed end, as shown in Fig. 5.4a. Those two vortices travel forward with a higher speed than that of the flame front and grow larger behind the flame front. The curvature radius increases gradually with the development of the flame as the vortices move on and get closer to the flame. In the meantime, the reverse flow becomes more intensive. Consequently, the velocity of the two side parts of the flame becomes higher than that of the middle section due to the reverse flow and the vortices. Then tulip formation is eventually initiated and promoted by this phenomenon. At the time $t = 9.6$ ms, the middle section just behind the flame front of the burned gas is completely dominated by the reverse flow. However, the velocity of the flow around the upper and lower section of the flame is almost forward directed as shown in Fig. 5.4b. The vortices continue moving forward after the tulip is formed for a short time, and the reverse flow dominates the center part of the burning area where the tulip cusp is located (Fig. 5.4c, d) so that the cusp propagates backward. With further propagation of the flame, the reverse flow spreads to a broader region. The whole region just behind the flame front is also dominated by the reverse flow, as shown in Fig. 5.4d. After the full formation of tulip flame, the vortices move backward (Fig. 5.4d), then decay and disappear gradually.

Some numerical works on the interactions among flame, induced reverse flow and vortices have been reported [6, 25]. However, further experimental works need to be conducted to validate the simulated results. Figure 5.5 shows the high-speed schlieren images of the motion of reverse flow and vortices initiated in the burned gas after tulip formation. These images experimentally demonstrate the motion of the reverse flow and the vortices in the burned gas. The experimental results confirm that the above numerical results are reasonable. Similar to the numerical observation, the reverse flow was first generated behind the flame prior to the vortices that appear subsequently near the sidewalls around $t = 12.3$ ms. The vortices travel backward and grow with the reverse flow. Meanwhile, the vortices have a tendency to move to the center of the duct and become disturbed gradually. After $t = 18.5$ ms, it is difficult to observe the reverse flow. This might be due to the decreasing of the velocity gradient in the burned gas. The breakup of the vortices after $t = 21$ ms implies that the induced flow instability grows more intensive. Note that the experimental reverse flow and vortices were not observed at the early stage. The main reason for this might be that the reverse flow and vortices in the initial stage of flame propagation are so weak that a sufficient velocity gradient, which can be recorded by the high-speed schlieren systems, could not be produced.

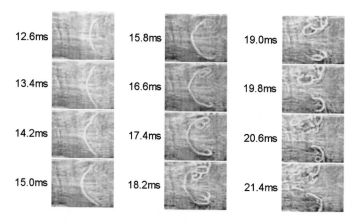

Fig. 5.5 High-speed Schlieren images of vortices and reverse flow near the flame front in the burned gas

The above experimental and numerical results demonstrate that the reverse flow and vortex motion near the flame front in the burnt region are important for the flame dynamics. It can be concluded that the flame front inversion and the actual formation of the tulip flame could be attributed to the interactions between the flame front, the flame-induced reverse flow, and the vortices.

5.4 Formation Mechanisms of Distorted Tulip Flames— Interactions Between Flame, Pressure Waves, and Flow

5.4.1 Interactions of Flame Front with the Vortex Motion in Burnt Gas

Basically, the expansion of the gas as it passes the flame front from the unburned to the burnt side initiates the flame-generated flow with a velocity difference $(E-1) \cdot S_L$. The interaction between the flame front and the flow field induced by combustion dominates the unsteady propagation. The interaction has been demonstrated in the formation of tulip flames [7, 8, 14, 18]. The dominant spatial scale of the fluid motion is usually tens of millimeters, which is of the order of the physical dimensions of the combustion vessel. According to Dunn-Rankin et al. [8], the flow field induced during flame propagation in a closed tube can be divided into three regions. The first region (far-field unburned region) is the unburned gas motion far from the flame front. In this region, the unburned gas motion is positive, irrotational and nearly one-dimensional, and the fresh gas is simply compressed by the burnt gas [7, 8, 18]. The second region (far-field burnt region) is the burnt gas motion far from the flame front. The burnt gas motion in this region is rotational in its nature [7, 8, 18]. The third region (near-field region) is the burnt and unburned gas motion near the flame front. The interaction between the flame front and the near-field gas motion leads to flame deformations, e.g., convex flames and tulip flames [8, 14].

In the present study, the main focus is put on the interaction between the flame front, pressure wave, and the flow in the third region that leads to the formation of the "distorted tulip" flame. It has been put into evidence both in experimental and numerical studies that vortices near the flame front are consistent with the onset of the tulip flame [7, 9, 14, 18] and can lead to the formation of tulip flames. This suggestion is fully supported by an analytical analysis [6, 19]. The interactions between flame front and pressure wave have been shown in Sect. 5.2.

Generally, a propagating flame front can be presented as a discontinuity with a jump of velocity normal to its surface. When the flame front is curved and oblique to the unburned gas, the velocity jump condition causes deflection of the flow direction. And the deflection phenomenon of the flow can create vortex motion in the close proximity of the flame front in the burnt gas [18]. Furthermore, a reverse flow will be induced behind the flame front during the formation of a tulip flame.

LES velocity field vectors and vortices induced by combustion behind the flame front during the formation of the "distorted tulip" flame are shown in Fig. 5.6. As the tulip flame grows further after the flame inversion, the reverse flow gradually dominates the near-field burnt gas behind the flame front and a small reverse flow appears in the unburned gas just within the confines of the tulip, as shown in Figs. 5.6a, b. This circumstance is similar to experimental results in [26, 27]. The reverse flow is the cause of the backward propagation of the original cusp, as shown in Fig. 3.15. After the complete formation of the classical tulip flame, the reverse flow begins to appear in the far-field unburned gas with the disappearance of the positive flow at around $t = 6.6138$ ms, as shown in Fig. 5.6a, as the pressure wave travels backward. Both the location of leading flame front and pressure dynamics reach a plateau period almost at the same time $t = 4.75$ ms, as shown in Figs. 3.14a and 3.16. Similarly the unburned gas has been entirely dominated by the reverse flow, as shown in Fig. 5.6b, just before the appearance of the circulating flow in the burnt gas.

The large-scale circulation starts to be created from the original tulip cusp behind the flame front at about $t = 7.0476$ ms, as shown in Fig. 5.6c. Just before this moment the unburned gas is dominated by the positive flow (from the left to the right) once again with the pressure wave propagating to the right. Almost at the same time, the flame front begins to accelerate again and the pressure starts to grow, as shown in Figs. 3.14b and 3.16. The original tulip cusp accelerates forward obviously after $t = 6.73$ ms (Fig. 3.15) as the intensity of the reverse flow behind the flame front is reduced, as shown in Fig. 5.6c. Before $t = 7.2203$ ms, the vortices have been well established just behind the primary tulip cusp which is surrounded by the vortices, as shown in Fig. 5.6d. At the same time, the flow in the proximity of the tulip cusp becomes positive, and the effects of this positive flow associated with the forward motion of the unburned gas within the tulip confines drive the primary cusp to accelerate forward further thus reducing its depth. And as the vortices grow, two small positive flow regions are created just behind the tips of the primary tulip cusps close to the sidewalls, as shown in Figs. 5.6d, e. This positive flow keeps the flame tip near the sidewall moving forward. However, behind the flame front, the flow near the center of the original tulip lips is always kept reverse due to the vortical motion, as shown in Figs. 5.6c–f. Therefore, the flame front decelerates near the center of the tulip lip. Finally, the "distorted tulip" flame is created and becomes more pronounced due to the effects of the motion of the vortices induced by the combustion, as shown in Figs. 5.6e–g.

The positive flow behind the original tulip cusp becomes weaker and ultimately changes into reverse flow again at $t = 7.6507$ ms, as shown in Fig. 5.6g. The reappearance of the reverse flow can maintain the tulip shape. The vortex center stays almost in the same position during the rapid process of the "distorted tulip" flame formation, as shown in Figs. 5.6d–g. Nevertheless, the flame front keeps traveling forward. As a result, the distance between the flame front and the vortex center becomes larger and the reverse flow displays a trend to dominate the region behind the entire flame front, as shown in Fig. 5.6g. The reverse flow becomes more drastic and moves in an opposite direction to the vortical motion just behind the

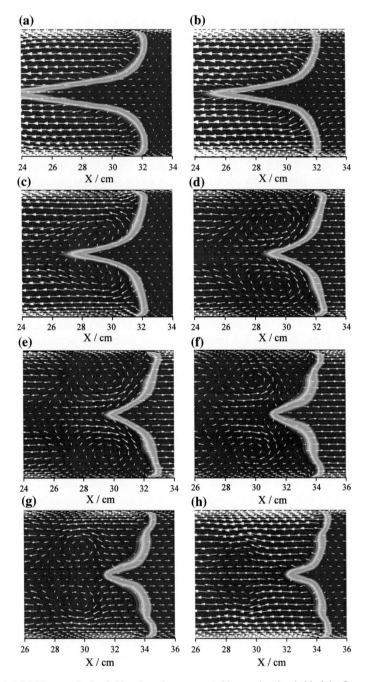

Fig. 5.6 LES Vector velocity field and vortices generated by combustion behind the flame front at the central plane of the duct at time instants $t = 6.6138$ ms (**a**), 6.8747 ms (**b**), 7.0476 ms (**c**), 7.2203 ms (**d**), 7.3061 ms (**e**), 7.5642 ms (**f**), 7.6507 ms (**g**), and 7.8235 ms (**h**)

original tulip cusp. The vortex begins to collapse and disappears due to the violent reverse flow around the cusp after the full formation of the "distorted tulip" flame, and the burnt region is entirely dominated by the strong reverse flow, as shown in Fig. 5.6h. The reverse flow reoccurs in the unburned gas at about $t = 7.8235$ ms (Fig. 5.6h), coinciding with another deceleration of the flame front and slowdown of pressure rise, as shown in Figs. 3.14 and 3.16, as the pressure wave travels backward once again.

Figure 5.7 shows the simulated pressure dynamics (grid 1) at three different locations in the LES simulations. P_1, P_2, and P_3 designate the pressures recorded at the bottom of the duct at $x = 0$ cm, 20 cm, and 40 cm from the ignition point, respectively. The pressure oscillations with half wave length equal to the length of the duct, initiated as the skirt of the flame touches the sidewalls, are seen in Fig. 5.7. During these oscillations, when the pressure maximum travels from right to left, a reverse flow is generated, e.g., at $t = 6.6138$ ms (t_1). This is followed by deceleration of the flame front (may be seen in Fig. 3.14) and the decrease of the rate of the pressure rise in the unburned region (P_3). Meantime, the pressure (P_1) in the burnt region starts to increase. The pressure oscillations continue to the end of the combustion process. After reaching the left end of the duct, the pressure wave is reflected and propagates in the direction of flame propagation again. The flow induced by the pressure wave interacts in the near-field burnt region with the flow of combustion products from the flame front, and this creates vortices behind the flame front. At the same time $t = 7.0476$ ms (t_2, Fig. 5.7), the unburned gas begins to move in the direction of flame propagation as shown in Fig. 5.6c. This is accompanied by the slowdown of the pressure rise in the burnt gas (P_1) and pressure growth in the unburned gas (P_3), as shown in Fig. 5.7. The reappearance of the reverse flow in the burnt gas behind the primary tulip cusp at $t = 7.6507$ ms (t_3), as shown in Fig. 5.6g, indicates that the velocity of the flow induced by the pressure wave decreases. The reverse flow reappears both in the unburned and burnt regions at about $t = 7.8235$ ms (t_4), as shown in Fig. 5.6h, and the pressure in the burnt

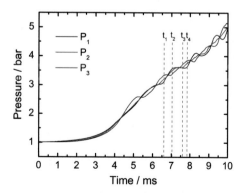

Fig. 5.7 Pressure dynamics in the LES simulation at $x = 0$ cm, 20 cm, and 40 cm. $t_1 = 6.6138$ ms (Fig. 5.6a), $t_2 = 7.0476$ ms (Fig. 5.6c), $t_3 = 7.6507$ ms (Fig. 5.6g), and $t_4 = 7.8235$ ms (Fig. 5.6h)

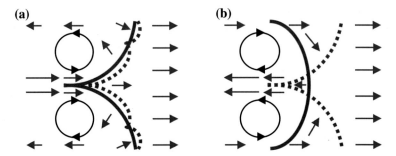

Fig. 5.8 Scheme showing the formation of: **a** "distorted tulip" flame, **b** classical tulip flame. *Solid* and *dashed lines* indicate the initial flame front shape and subsequent flame shape. *Arrows* indicate a characteristic flow velocity field

region (P_1) is growing fast once again, as shown in Fig. 5.7. The period of the oscillations becomes shorter as the flame propagates due to the increase of the speed of sound with temperature rise both in the unburned and burnt gas and growth of the duct volume occupied by combustion products.

The interaction of the flame front with the combustion-induced flow and the formation of the "distorted tulip" flame in comparison with the generation of a classical tulip shape are shown schematically in Fig. 5.8. It can be seen that a counterclockwise vortex is created just behind the upper lip of the tulip flame, while a clockwise vortex is created just behind the lower lip during the formation of the "distorted tulip" flame. However, the direction of vortices in the classical tulip flame (Fig. 5.8b) is contrary to the counterparts in the "distorted tulip" flame (Fig. 5.8a).

The above analysis shows that the interaction between the flame front, pressure wave, and the induced flow plays an important role in the flame dynamics and the "distorted tulip" flame formation. The effects of the vortices generated in the proximity of the flame front lead to the continuous change of the flow field, causing the flame to propagate faster at the sidewalls and at the original tulip cusp compared with flame propagation near the center of the original tulip lips. Consequently, the "distorted tulip" flame is formed due to the difference in flame propagation rates.

5.4.2 Taylor Instabilities

Figure 5.9 presents the propagation speed (displacement speed) of the flame leading tip and the pressure growth rate inside the chamber as a function of time (hydrogen concentration 30 % by volume). Before flame inversion, the flame front at the centerline of the duct is taken as the flame leading tip while after inversion the tip of the upper part of the flame front is treated as the flame leading tip. The propagation speed of the flame leading tip is calculated based on the changes in its position

Fig. 5.9 Experimental
Propagation speed of the
flame leading tip and pressure
growth rate inside the duct

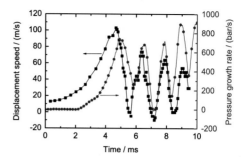

(distance from the ignition point) with time. As indicated above, the flame front propagates with its speed oscillating periodically, and the initiation of both the tulip and distorted tulip flame is consistent with the sudden deceleration both of flame front and pressure growth. The first deceleration is caused by the violent reduction in the flame surface area as the flame touches the sidewalls. It has been experimentally, theoretically, and numerically put into evidence that the sudden flame deceleration subsequent to the first contact with the sidewalls of the duct results from the rapid decrease of flame surface area [3, 13]. An example of quantitative evolution of flame surface area versus time in tulip flame propagation may be found in [3]. Additionally, vortex motion is commonly produced in the burnt region after the flame has touched the sidewalls. The recirculation can advect the center part of the flame front backward later on, thus creating a cusp. Therefore, the vortical motion in the burned gas could further the deceleration of the flame leading tip during flame inversion process. Nevertheless, the mechanism resulting in the later decelerations is different. Actually, it appears that the distorted tulip flame is initiated without significant reduction in the flame surface area, as shown in Fig. 5.9 ($t = 7.0$ ms). And the above numerical simulation indicates that the deceleration is achieved prior to the creation of vortex motion during the formation of distorted tulip flame. The drastic flame deceleration implies that a strong decrease of the flow velocity in the unburned gas is caused.

It has been known that oscillations of pressure growth gate are triggered by the sudden deceleration of the flame front at the third stage. The pressure growth rate oscillates periodically in phase with the flame propagation speed. A pressure wave is triggered by the contact of flame front with the lateral walls and travels forth and back inside the duct. The time interval between the first distorted tulip flame and the second one is about 1.33 ms. The period of the pressure wave deduced from the pressure growth rate is around 1.30 ms, approximately equal to the interval of the two distorted tulip flames. The evolution of the flame shape, flame propagation speed, and pressure growth rate reveals that the interaction between the distorted tulip flame and the pressure wave is strong. Gonzalez [23] numerically demonstrated that the periodically oscillating behavior of a premixed flame propagating in a closed tube at the later stage results from the interaction between flame and pressure waves. And the periodic flame acceleration/deceleration drives the flame to display an oscillating cellular pattern, which is akin to Taylor instability. According to the experiments by Markstein [5], sudden deceleration of a flame front can lead to flame inversion that is similar to the tulip phenomenon. In his experiments, a curved flame is indented in the center after a shock wave passes the flame front in the direction from the unburned mixture (heavier) to the burnt region (lighter). On the basis of Taylor instability theory, Markstein [5] suggested that this flame inversion is caused by the velocity field (drastic velocity decrease) behind the shock wave as it passes the flame. Although there is no shock wave in the present combustion process, the physical situation during the distorted tulip flame formation in this study is very similar to that in the Markstein's experiments. First, a drastic change in the unburned gas velocity with the onset of the distorted tulip shape is produced by the pressure wave effect, causing violent velocity decrease and reverse flow in the unburned region, as indicated in Sect. 5.4.1. The pressure wave in this study acts as a source similar to a shock wave, leading to velocity change in the unburned mixture. Second, the distortions are initiated immediately the flame undergoes a sudden deceleration and two secondary tulip cusps are consequently created on the primary tulip lips. Markstein observed the velocity change caused by a shock of pressure ratio 1.3 for a stoichiometric butane/air flame in a 7.6-cm-square-closed tube is 97.5 m/s. The flame propagation speed changes accompanied by the first and second distorted tulip flames are 73.7 and 63.8 m/s in the present case. And the numerical study by Gonzalez [23] shows that the velocity changes caused by pressure wave in a closed duct with aspect ratio of six are 40–50 times the laminar burning velocity of hydrocarbon. Thus the resulting velocity changes in the present experiment are comparable with those in the experiments of Markstein and numerical simulation of Gonzalez. In the meantime, the pressure oscillations in this work are pretty similar to those at the early flame stage in the numerical simulation of Gonzalez. Therefore, it is reasonable to conclude that the distorted tulip flame have the same physical origin as that in the experiments of Markstein [5].

It should be noted that the distance between the flame leading tip and the location of pressure transducer decreases as the flame propagates since the transducer is at a fixed location 40 cm from the ignition point in horizontal direction. The triggering time of the first oscillation of the pressure growth rate is larger than that of the flame tip propagation speed. A time delay of about 0.18 ms between the first oscillations of pressure growth rate and flame tip speed is obtained from Fig. 5.9. The distance between the flame tip and the pressure transducer is around 7.6 cm as the pressure wave is triggered. And the calculated time delay is 0.19 ms by assuming the sound speed in the hydrogen/air mixture to be 400 m/s. It can be seen that the calculated time delay is very close to that derived from Fig. 5.9. On the contrary, the triggering time of the second oscillation of the pressure growth rate is slightly smaller than that of the tip speed since the pressure wave is reflected from the right end of the duct. And the second time delay is shorter than the first one. This can be explained by the fact that the distance between the flame tip and the pressure transducer becomes smaller as the flame develops further while the sound speed increases with the temperature rise.

In conclusion, the interaction between flame front and the pressure wave triggered by the first contact of flame with the sidewalls of the duct is strong. The pressure wave leads the flame to decelerate periodically. And the flame propagation speed oscillates in phase with the pressure growth rate. The sudden deceleration of the flame front drives the flame front to display a distorted tulip shape. The dominant physical process involved in the formation of distorted tulip flame is quite similar to that in the experiment of Markstein [5]. The formation of the distorted tulip flame could be a result of Taylor instability.

5.5 Summary

(1) The pressure wave, triggered by the contact of the flame with the duct sidewalls, plays a dominant role in the formation of the distorted tulip flame. The flame propagation speed oscillates in phase with both the pressure growth rate and flow velocity in the unburned mixture. The significant interaction of the flame with the pressure wave is responsible for the periodical flame deceleration and the consequent distorted tulip flame formation. Both the experiment and numerical calculation show that the physical process in the distorted tulip formation could be the same as that in the interaction of flame front with shock wave in the experiments of Markstein [5].

(2) The interactions between the flame front, the flame-induced reverse flow, and the vortices directly cause the inversion of the flame front and the subsequent formation of the tulip flame. The reverse flow is formed in the center region of the burnt zone prior to the formation of vortices. The vortices start to appear near the left closed end. The reverse flow becomes stronger and spreads to the cusp zone and to the entire burnt region just behind the flame front. Two vortices are formed near the two sidewalls and overtake the flame front. Consequently, the propagation speed of the flame front near the sidewalls becomes higher than that of the center region. This discrepancy in flame propagation speeds leads to the formation of the tulip flame. The two vortices grow large and break up finally with more flow instabilities induced.

(3) The interactions between flame front, pressure wave, and combustion-generated flow were examined in detail in the LES simulations. The formation mechanism of distorted tulip flame was suggested. The close correspondence between the numerical and experimental results indicates that the circulation near the flame front in the burnt gas is the dominant physical process involved in the formation of the distorted tulip flame. The effects of the vortices generated in the burnt gas behind the tulip flame front (due to interactions between the flame front, acoustic waves, and combustion-generated flow) create the conditions required for the production of different flame propagation rates, which then lead to the formation of "distorted tulip" flame. In addition, the flow field during the "distorted tulip" flame formation undergoes complex changes around the flame front.

(4) The theoretical analysis showed that Taylor instability could play an important role in the formation of distorted tulip flame. The formation of the distorted tulip flame can be identified as a manifestation of Taylor instability.

References

1. Ellis OC de C (1928) Flame movement in gaseous explosive mixtures. Fuel Sci 7:502–508
2. Salamandra GD, Bazhenova TV, Naboko IM (1959) Formation of detonation wave during combustion of gas in combustion tube. Proc Combust Inst 7:851–855
3. Bychkov V, Akkerman V, Fru G, Petchenko A, Eriksson LE (2007) Flame acceleration in the early stages of burning in tubes. Combust Flame 150:263–276
4. Bychkov VV, Golberg SM, Liberman MA, Eriksson LE (1996) Propagation of curved stationary flames in tubes. Phys Rev E 54:3713–3724
5. Markstein GH (1956) A shock-tube study of flame front-pressure wave interaction. Proc Combust Inst 6:387–398
6. Matalon M, Metzener P (1997) The propagation of premixed flames in closed tubes. J Fluid Mech 336:331–350
7. Starke R, Roth P (1986) An experimental investigation of flame behavior during cylindrical vessel explosions. Combust Flame 66:249–259
8. Dunn-Rankin D, Barr PK, Sawyer RF (1986) Numerical and experimental study of "tulip" flame formation in a closed vessel. Proc Combust Inst 21:1291–1301
9. Dunn-Rankin D, Sawyer RF (1998) Tulip flames: changes in shape of premixed flames propagating in closed tubes. Exp Fluids 24:130–140
10. Gonzalez M, Borghi R, Saouab A (1992) Interaction of a flame front with its self-generated flow in an enclosure—the tulip flame phenomenon. Combust Flame 88:201–220
11. Guenoche H (1964) Flame propagation in tubes and in closed vessels. In: Markstein GH (ed) Nonsteady flame propagation. Pergamon Press, New York, p 107
12. Matalon M, Mcgreevy JL (1994) The initial development of a tulip flame. Proc Combust Inst 25:1407–1413
13. Clanet C, Searby G (1996) On the "tulip flame" phenomenon. Combust Flame 105:225–238
14. Rotman DA, Oppenheim AK (1988) Aerothermodynamic properties of stretched flames in enclosures. Proc Combust Inst 21:1303–1312
15. Dold JW, Joulin G (1995) An evolution equation modeling inversion of tulip flames. Combust Flame 100:450–456
16. Nkonga B, Fernandez G, Guillard H, Larrouturou B (1993) Numerical investigations of the tulip flame instability—comparisons with experimental results. Combust Sci Technol 87:69–89
17. Marra FS, Continillo G (1996) Numerical study of premixed laminar flame propagation in a closed tube with a full Navier-Stokes approach. Proc Combust Inst 26:907–913
18. Dunn-Rankin D, Sawyer RE (1985) Interaction of a laminar flame with its self-generated flow during constant volume combustion. In: Proceedings of the 10th ICDERS, Berkley, California, F August
19. Metzener P, Matalon M (2001) Premixed flames in closed cylindrical tubes. Combust Theor Model 5:463–483
20. Kaltayev AK, Riedel UR, Warnatz J (2000) The hydrodynamic structure of a methane-air tulip flame. Combust Sci Technol 158:53–69
21. Pocheau A, Kwon CW (1989) Proceedings of A.R.C. colloquium. C.N.R.S.-P.I.R.S.E.M., Paris, p 62
22. Matalon M (2009) Flame dynamics. Proc Combust Inst 32:57–82

23. Gonzalez M (1996) Acoustic instability of a premixed flame propagating in a tube. Combust Flame 107:245–259
24. Chomiak J, Zhou G (1996) A numerical study of large amplitude baroclinic instabilities of flames. Proc Combust Inst 26:883–889
25. Zhou B, Sobiesiak A, Quan P (2006) Flame behavior and flame-induced flow in a closed rectangular duct with a 90 degrees bend. Int J Therm Sci 45:457–474
26. Liberman M (2003) Flame, detonation, explosion—when, where and how they occur. In: Third international disposal conference. Karlskoga, Sweden, pp 5–23
27. Verhelst S, Wallner T (2009) Hydrogen-fueled internal combustion engines. Prog Energy Combust Sci 35:490–527

Chapter 6
Conclusions and Further Work

6.1 Summary

The dynamics and mechanisms of premixed hydrogen–air flames propagating in ducts have been studied in this thesis. The primary focus of this thesis has been to reveal flame behaviors, characteristics, and the physical origins dominating the flame evolution. The reliability and accuracy of numerical and theoretical models and methods for transient hydrogen–air flame propagation have also been examined. This work may provide a basic understanding of the fundamentals and applications of combustion and explosion in confined regions related to the safety of hydrogen as an energy carrier.

In the experiments, high-speed schlieren photography and pressure records were used to investigate the behaviors and characteristics of premixed hydrogen–air flames propagating in ducts under various conditions, such as the flame shape and position changes as a function of time, pressure dynamics, and interactions between flame and pressure waves. In the numerical simulations, both a dynamically thickened flame model with a laminar flow model and a burning velocity model with a LES model were used to simulate the premixed hydrogen–air flame propagation and then to explain the inerations between flame front, combustion-generated flow, and pressure waves. The CFD numerical models were validated against the experimental and theoretical results. On the basis of experimental and numerical results, an analytical study was performed and a theoretical model of premixed hydrogen-air flame propagation in a closed tube was suggested.

6.2 Main Conclusions

The main conclusions of this work can be drawn as follows:

(1) Experiments of premixed hydrogen–air Flame propagation in ducts

© Springer-Verlag Berlin Heidelberg 2016
H. Xiao, *Experimental and Numerical Study of Dynamics
of Premixed Hydrogen-Air Flames Propagating in Ducts*,
Springer Theses, DOI 10.1007/978-3-662-48379-4_6

The premixed hydrogen–air flames evolving in the ducts undergo a series of shape and structure changes. One of the outstanding findings is that a salient distorted tulip flame is formed after the formation of a well-pronounced tulip flame in the equivalence ratio range of $0.84 \leq \Phi \leq 4.22$ in the closed duct. A second distorted tulip flame can be produced just before the disappearance of the first one. Classical tulip flame forms in a large range of equivalence ratio, i.e., $1.17 \leq \Phi \leq 4.05$ in the half-open duct and $0.49 \leq \Phi \leq 7.14$ in the closed duct, respectively. The onset of flame deformations, e.g., tulip and distorted tulip flames, coincides with the deceleration both of pressure rise and flame propagation (displacement) speed. The dynamics of a distorted tulip flame is different from that of a classical tulip flame. A distorted tulip flame experiences more complex shape changes and more unstable combustion processes than a classical tulip flame.

The formation of flame shape changes greatly depends on the mixture composition. In addition, the equivalence ratio range that can forms tulip flame is much wider in the closed duct ($0.49 \leq \Phi \leq 7.14$) than that in half-open duct ($1.17 \leq \Phi \leq 4.05$). The dimensionless formation time of quasi-plane/plane flame exponentially decreases as the equivalence ratio increases both in the half-open and closed ducts. And the tulip flame formation time in the half-open duct is nearly equal to that in the closed duct at the same equivalence ratio. The formation position of a quasi-plane/plane flame has an approximately negative correlation with the expansion ratio.

Gravity has an effect to some extent on the flame dynamics. Gravity can lead the tulip flame to collapse at low equivalence ratios in a different way from that at high equivalence ratios. Nevertheless, it may not result in substantial difference in the tulip flame formation. Opening ratio has a significant impact on the flame dynamics in a partially opened duct. Smaller opening ratio leads to more drastic flame shape changes. When the opening ratio is in the range of $\sigma \leq 0.4$, a noticeably distorted tulip flame can form subsequent to a well-established classical tulip flame. The propagation speed of flame leading tip increases with the increase of opening ratio. Both the pressure growth rate and its oscillation amplitude in the duct increase with decreasing the opening ratio.

(2) Numerical simulations of dynamics of premixed hydrogen–air flames propagating in ducts

The flame propagation was simulated as 2D and 3D chemically reacting flows. Either TF model or LES combustion model was employed in the numerical simulations. Detailed experimental results were first shown for comparisons of numerical simulations with experiments. For the near-stoichiometric (hydrogen concentration 30 % by volume) hydrogen–air flame, five stages of flame dynamics have been proposed in the distorted tulip flame propagation, i.e., spherical flame, finger-shaped flame, flame with skirt touching sidewalls, tulip flame, and distorted tulip flame. The distorted tulip flame deforms into a salient "triple tulip" shape as the secondary tulip cusps approach the center of the primary tulip lips. The distorted tulip flame repeatedly undergoes collapse and reappearance processes. Both the flame tip propagation (displacement) speed and pressure rise show periodic oscillations during the propagation of distorted tulip flame. The formation of the distorted tulip flame coincides with the sudden decrease in both the velocity of the leading flame front and the rate of pressure rise.

In the 2D CFD simulations, a dynamically thickened flame model has been used to account for the premixed combustion. The dynamics of the premixed flame with distorted tulip shapes, observed in the experiments, has been reasonably reproduced in the numerical simulations. The TF technique associated with the 19-step chemical reaction scheme was shown to be pretty reliable for studying the premixed hydrogen–air flame propagation and interaction between flame front and pressure waves in the closed vessel. In the 3D numerical simulations with a laminar solver, TF model was also used. The experimental results were used to assess the capability of the numerical models for calculating the premixed combustion process. The satisfactory agreement between the 3D numerical simulations and experiments indicates that the TF model with the seven-step chemistry scheme is also reliable for predicting the transient premixed hydrogen–air combustion in the closed duct. The flame behavior and pressure dynamics have been well reproduced in the numerical simulations. Oscillations occur in the flame front dynamics and the pressure rise both in the experiment and numerical simulation after the flame reaches the lateral walls. Both the flame propagation speed and pressure build-up were underestimated after the formation of tulip flame in the numerical simulation since the flame wrinkling effects blow grid resolution were unresolved by the TF model.

The multiphenomena combustion model, a type of burning velocity model, has been applied in the LES computations. The LES model reproduced the experimental observations with reasonable accuracy with sufficient grid resolution. The good agreement between the LES simulations and experiments confirms that the multiphenomena combustion model provides a reasonable prediction of the premixed hydrogen–air combustion in the closed duct. The five typical stages of flame propagation, as observed in the experiment, were captured in the simulations run on the finer grid. The grid resolution has an important influence on the LES simulation. The simulations on the course grid could reproduce the classical tulip flame, but not the "distorted tulip" flame. As expected, the classical tulip flame on the coarse grid is less pronounced due to the excessively thickened flame front. Furthermore, the pressure build-up rate obtained on the coarse grid were underestimated as well.

(3) Theoretical analysis of premixed hydrogen–air flame propagation in ducts

The theoretical analysis was based on previous peers' analytical studies, experiments, and LES simulations in this work. The five stages of the dynamics of distorted tulip flame were examined. A theoretical model was preliminarily suggested by analyzing the interaction between flame front and pressure wave. Agreement between the theory, experiment, and LES simulations at the early stage of the flame propagation is satisfactory.

It was put into evidence that the SGS wrinkling effects play an important role in flame acceleration, whereas the turbulence in the unburned gas is weak and of minor importance for the flame propagation. The major contribution of wrinkling effect is from the self-turbulization of flame front. The flow was found to be nearly laminar, but it is necessary to model the effects of various physical phenomena that enhance the burning rate.

The study showed that both the tulip and distorted tulip flames can form without wall friction. This indicates that wall friction is unimportant for the formation of both tulip and distorted tulip flame in the duct. The interaction between flame front and pressure waves can be substantially independent of boundary condition at the sidewalls, although boundary layer has an effect on the flame propagation speed close to the sidewalls.

(4) Mechanisms of flame deformations in premixed hydrogen–air flame propagation

The interactions between the flame front, the flame-induced reverse flow, and the vortices directly cause the inversion of the flame front and the subsequent formation of classical tulip flame. Due to the presence of both the reverse flow and the vortical flow, the propagation speed of the flame front in the proximity of the duct sidewalls is higher than that of the center region. This leads the tulip flame to form.

Phenomenologically, the interactions between flame front, pressure wave, and combustion-generated flow are essential for the formation of distorted tulip flame. The close correspondence between the LES and experimental results suggests that the vortex motion near the flame front in the burnt gas is the actual physical process resulting in the formation of distorted tulip flame. The effects of the vortices that were generated in the burnt gas behind the tulip flame front create the conditions favorable for the formation of distorted tulip flame.

It was found that the pressure wave, triggered by the contact of the flame with the duct sidewalls, is dominant in the formation of distorted tulip flame. The interaction of flame with pressure waves causes the periodic oscillating behavior of flame. The effect of pressure wave is responsible for the flame deceleration and the consequent distorted tulip flame formation. Both the experiments and numerical calculations show that the physical origin in the distorted tulip formation is the same as that in the interaction of flame front with shock wave in the experiments of Markstein [1]. The theoretical analysis showed that the formation of distorted tulip flame can be identified as a manifestation of Taylor instability.

6.3   Future Research

The present study has contributed to an increased knowledge of premixed hydrogen–air fame propagation in tubes. Some general recommendations for future research are listed below:

(1) Nonlinear development of flame instabilities such as DL and RT instabilities plays an essential role in the actual evolution of premixed flame front. The flame shape and speed are significantly influenced by the nonlinear interactions of flame front with flow and pressure waves. It is necessary to conduct further theoretical and numerical studies of the nonlinear development of the unstable flames propagating in tubes.

(2) With respect to thickened flame model, the flame wrinkling effects at scales under grid resolution are important for accurately predicting the flame

dynamics in a tube, especially in the later flame stages when the hydrodynamic and combustion instabilities become significant. Further work needs to be performed, and there is a requirement for the modeling of the wrinkling effects in order to allow a more quantitative prediction of the premixed hydrogen–air flame propagation.

(3) The moderate-scale experiments and numerical simulations of premixed hydrogen–air flame propagation in ducts were performed in this work. From the point of view of hydrogen safety engineering, large-scale experiments and calculations of hydrogen–air explosions can be of realistic importance. In larger-scale hydrogen–air flame propagation, laminar-to-turbulent flame transition can be one of the crucial features of the explosion dynamics. On the other hand, the explosion in industries is usually a large-scale, unsteady and high-speed combustion process. To provide basic data and theories of practical explosions, it is necessary to carry out large-scale experiments and numerical simulations of hydrogen explosions in air under a variety of conditions. Turbulent combustion is crucial for flame acceleration and detonation initiation, but the understanding of turbulent flame is incomplete. Therefore, study of turbulent flame in hydrogen–air flame propagation is also of scientific and engineering interest.

(4) Deflagration-to-detonation (DDT) remains one of the major unsolved problems in combustion and explosion research field. DDT also has immensely important applications ranging from safety engineering to propulsion engineering and to astrophysics and cosmology. Future work can be focused on the flame acceleration and DDT in hydrogen–air mixtures in confined regions. In addition, obstacles are generally present during flame turbulization and DDT processes in industrial explosions, such as coal mine explosions. Thus the effects of obstacles on flame acceleration and DDT can also be planned for future study.

(5) Hydrogen as an energy carrier is one of the most promising alternative fuels in the future. The hydrogen safety issues concern fires and explosions in refueling stations, tunnels, garages, gas pipelines, etc. The practical prevention and mitigation of hydrogen explosions in these hazardous industrial scenarios rely on the development of safety science, technologies, standards, specifications, and regulations of hydrogen safety. Therefore, it is of great importance to provide basic understanding of hydrogen explosions in terms of engineering science that can help us develop technologies and standards for hydrogen safety related to explosions in confined regions.

Reference

1. Markstein GH (1956) A shock-tube study of flame front-pressure wave interaction. Proc Combust Inst 6:387–398

Printed in the United States
By Bookmasters